About Island Press

Since 1984, the nonprofit organization Island Press has been stimulating, shaping, and communicating ideas that are essential for solving environmental problems worldwide. With more than 1,000 titles in print and some 30 new releases each year, we are the nation's leading publisher on environmental issues. We identify innovative thinkers and emerging trends in the environmental field. We work with world-renowned experts and authors to develop cross-disciplinary solutions to environmental challenges.

Island Press designs and executes educational campaigns, in conjunction with our authors, to communicate their critical messages in print, in person, and online using the latest technologies, innovative programs, and the media. Our goal is to reach targeted audiences—scientists, policy makers, environmental advocates, urban planners, the media, and concerned citizens—with information that can be used to create the framework for long-term ecological health and human well-being.

Island Press gratefully acknowledges major support from The Bobolink Foundation, Caldera Foundation, The Curtis and Edith Munson Foundation, The Forrest C. and Frances H. Lattner Foundation, The JPB Foundation, The Kresge Foundation, The Summit Charitable Foundation, Inc., and many other generous organizations and individuals.

The opinions expressed in this book are those of the author(s) and do not necessarily reflect the views of our supporters.

Multisolving

Multisolving

CREATING SYSTEMS CHANGE
IN A FRACTURED WORLD

Elizabeth Sawin

ISLANDPRESS | Washington | Covelo

Design and Typesetting by 2K/DENMARK.

Sustainable Typesetting® ID: ST49535052-24-B02
The ID number certifies that this book fulfills the standards and recommendations of Sustainable Typesetting® providing reduced CO_2 emissions.

Library of Congress Control Number: 2024935669

All Island Press books are printed on environmentally responsible materials.

Manufactured in the United States of America
10 9 8 7 6 5 4 3 2 1

Keywords: adaptability; behavior shift; climate change; complex systems; cross-sectoral networks; destabilization; diversity; Donella Meadows; ecological restoration; environmental and social crises; environmental justice; equity; feedback loops; interdisciplinary solutions; living systems; multiracial coalitions; racial justice; redundancy; resilience; resistance to change; stocks and flows; sustainability; systems thinking; worldviews

*To multisolvers everywhere, and
to Phil, Jenna, and Nora, in
love and appreciation.*

Contents

Converging Crises, Cascading Solutions

Change, emergencies, even crises, are normal occurrences in our lives. We build and engage with systems to help us prepare for, and cope with, disruption. As individuals we insure our homes and cars, learn first aid, and maybe even volunteer for the local fire department. We practice conflict mediation and pay taxes to help support emergency responsiveness. We save money for a rainy day and donate to the local food pantry to help others who have fallen on hard times.

Shouldn't that be enough?

With regard to future crises, shouldn't it be enough to do what previous generations have done? Shouldn't it be enough to take some personal precautions and create institutions to do the rest? Well, if you've reached for this book, then the odds are you suspect that the answer to both questions is no. Are you seeing signs of increasing disruption? Are you witnessing problems cascade, converge, and amplify one another? Are you feeling as though maybe something more is needed in response?

Climate change is powering stronger storms and more intense wildfires.[1] Biodiversity is falling, and the frequency of economic disruptions is quickening.[2]

In many places democracy is weakening, and in others the elite are consolidating wealth.[3] In the face of these mutually reinforcing crises, we need different strategies for coping. The old ones seem to be proving not up to the job.

There is no one-size-fits-all experience of destabilization. For wealthy North Americans or Europeans, the shocks may feel muffled. For a poor family in the Global South, they may be cataclysmic. But if it feels to you that instability is rising and that the periods of calm between storms are shortening, you are not alone. On a planet with growing ecological crises, the trend is toward more destabilization rather than less.

When I wrote the following words, in notes for what eventually became this book, in mid-2020 in the United States, this is what destabilization looked like from my vantage point:

> COVID-19 is spreading unchecked in some states, and hospitals are being overwhelmed. The coronavirus is infecting and killing people of color at a higher rate than white people.[4] This is at least partly due to inequities in environmental health, access to care, and access to remote jobs. For over a month, protestors have been in the streets of most cities in the US in response to police violence directed at Black men. The protestors are calling for the abolition of police and prisons. Schools closed in the spring, and many summer activities have been canceled or moved online. Parents are stretched thin. Meanwhile, most of my country is in the midst of a heat wave that is already two weeks old and is expected to last at least two more weeks. The pattern, called a "heat dome" is made more likely and more intense by climate change.[5]

By the time this book is in your hands the specific symptoms of global and local instability will be different. But if trends continue, one or perhaps both of us may still be navigating some flavor of it.

Not only are we experiencing increased instability, sometimes it seems to be feeding on itself. Notes to myself from early in the pandemic captured how destabilization in one place can ripple out to cause destabilization elsewhere:

> This month roughly one-third of households in the United States missed paying their rent or mortgage.[6] Some analysts are predicting a wave of evictions and utility shutoffs in the midst of the heat wave.[7] Where will evicted people go during a heat wave, people ask? Traditional solutions—cooling centers, libraries—are closed due to the pandemic. Evictions combine with heat waves to collide with a pandemic to create a situation that exceeds the system's ability to cope.
>
> Many of those who still have jobs are working from home, in the midst of the heat wave. Without air-conditioned office buildings to work from, many will purchase air conditioners. Energy use will increase, and so will climate-changing

greenhouse gas emissions. The solution to a heat wave, which was made worse by climate change, could make future climate change even worse.

And so it goes. Problems escalate, converge, and cascade. They shape-shift and spread. Here are some of the patterns I'm noticing. You may recognize them, and others, in your own life.

Sometimes, crises arrive faster than we can cope with them. Given enough time, we could manage many of the types of crises we are currently facing. But when they arrive too fast or in dangerous combinations, they overwhelm our ability to cope. In 2020 for instance, the city of Lake Charles, Louisiana, had yet to recover from Hurricane Laura when Hurricane Delta struck.[8] Houses whose roofs were still covered with blue tarps from Laura's impact were even more vulnerable when Delta arrived. The radar system, damaged by Laura, was not working optimally to give people information about Delta's approach.[9] That's what happens when crises arise faster than the speed of recovery.

In addition to the same types of crises following one after the next with increasing frequency, communities or businesses may also experience several *different* types of crises at close to the same time. When crises are slow and sporadic, responding to several types of crises using the same coping capacity can work. It is efficient, in fact. But what happens when the same emergency response department has multiple overlapping crises to deal with?

As the 2020 hurricane season began in North America, I was on a conference call with officials worried about the hurricane readiness of a major US city. The city had depleted its emergency response budget trying to manage COVID-19. As a result, the whole city stood vulnerable at the beginning of hurricane season. One crisis had depleted its capacity to cope with other potential crises.

Some crises feed on themselves; others erode our ability to cope. Viral pandemics—including COVID-19, of course—are classic examples of crises that feed on themselves. All else being equal, the more people who are infected in a community, the more people are exposed, leading to a rising tide of infections, sometimes with devastating results for health systems and for human well-being.

Even if a crisis doesn't feed on itself, it can cause enough damage that it becomes harder to respond to the next crisis. When a house is burned out by wildfire and not rebuilt, that's one fewer property tax payer in a community. It's also a smaller budget with which to respond to future fires. If a business lays off workers when supply chain issues disrupt the production schedule, it will take longer to fix mechanical problems that crop up on the production line because the most knowledgeable workers won't be on the floor. Countries that have spent heavily to cope with the pandemic have fewer resources to invest in climate change adaptation. Each of these is an example of one crisis depleting the ability to respond to others.

Solutions to one problem can make others worse. In 2018 sparks from electric transmission lines in California led to climate change–exacerbated fires that burned thousands of acres of land and destroyed homes and businesses.[10] In response, for the following fire season, one electric utility, Pacific Gas and Electric Company (PG&E), instituted power shutoffs. The shutoffs may have prevented some fires, but they also caused small businesses to lose revenue. And people who lost the contents of their refrigerators and freezers were put at risk of food insecurity. The way the utility addressed the fire problem contributed to new problems.

Vernice Miller-Travis, an environmental justice leader in the United States, wrote in 2020 about the "synergistic epidemic" of COVID-19 and environmental injustice, another example of complex, interacting, mutually exacerbating problems:

> While most Americans are confronting the coronavirus pandemic, communities of color are confronting something worse, the Syndemic of Coronavirus and Environmental Injustice. A syndemic is a synergistic epidemic. It is a set of linked health problems contributing to excess disease. To prevent a syndemic, one must control not only each affliction in a population but also the forces that tie those burdens together.
>
> Constant exposure to high levels of air toxics in communities of color has already resulted in explosive levels of respiratory disease, including asthma, chronic obstructive pulmonary disease, and emphysema, as well as heart disease. These pre-existing conditions have compounded the devastating impact of this pandemic; communities of color are now experiencing the highest rates of infection and death from COVID-19 in the United States. Lax attention to poor air quality has provided the perfect conditions for coronavirus to ravage minority neighborhoods.[11]

Failure to address one crisis diminishes gains made on others. In an interconnected world with long, interacting chains of cause and effect, sometimes the efforts to solve a problem in one sector are successful, only for some other neglected problem to feed back around and erode the initial progress.

For instance, let's say that new housing is designed to be energy efficient. Terrific! This will allow the residents to save more of their paycheck for other expenses, like food, education, and medicine. But if the city where the new housing sits hasn't mustered a strong climate change adaptation program, the combination of urban heat island effect and longer and more frequent periods of extreme heat might create a stronger need for air-conditioning in the summer. Energy bills go back up, and residents' savings go down.

And so on. When problems interconnect, focusing on them one at a time can result in unanticipated and sometimes devastating backlashes.

Crises grow in size and scale because vested interests resist change. Some problems, whose consequences are extremely dangerous, have well-documented solutions that are waiting to be implemented. Biodiversity loss and climate change come to mind. For crises like these, the scientific consensus is strong. However, a few powerful interests stand to lose a lot if action to solve the problem is taken, and so they deny the problem and prevent the strong response that is needed. There is a mismatch in power between those working for change and those who resist it, and that delays the response to the crisis.

———❧———

When you look at these patterns of crises and responses, though they occur at various scales, in a range of geographies, and concern different issues, what do you see?

I see problems that are not yielding to the standard ways we have been approaching them.

I see problems that are interconnected and solutions that are too often siloed.

I see short- and long-term components being addressed with solutions that focus on only one timescale.

I see problems that span silos and people working within silos trying to solve them.

And when I talk to the people who are trying to address these problems, I hear frustration, exhaustion, fear, and demoralization.

I see, more than anything, problems that cry out for different approaches.

The good news is people are experimenting with different approaches around the world, designing weblike solutions for weblike problems. The *very* good news is that we can all learn from these experiments and add to them. It's not too late to start. But there's also not a moment to lose.

The term I use for these weblike solutions is *multisolving* and, of course, I will have much more to say about multisolving in the chapters that follow. But I haven't always seen the potential of these silo-crossing approaches; I haven't always understood just how much difference multisolving could make in our world. In fact, for much of my career, I was far from being a multisolver.

My work focused quite narrowly on a single concern: global climate change and the greenhouse gas emissions that contribute to it. That was good, important work, and I'm glad I had the opportunity to do it. I taught about the carbon cycle and how the extraction and burning of fossil fuels push it out of balance. I attended United Nations climate conferences. I watched with pride as our think tank's analysis was shared on the front pages of newspapers. I celebrated each time our tools informed the thinking of top government officials and grassroots climate activists. All this work had a single purpose: squeezing greenhouse gases from the global economy. All of it was satisfying and all of it was needed. It still is. But … none of it was sufficient. Every year of my career, the greenhouse gas emission levels we and so many others were working to drive downward continued instead to climb.

Amid one particularly disappointing round of UN climate negotiations, I pulled a negotiator from Latin America aside. What could we add to our analysis, I asked her, that could help to unlock larger climate commitments from the UN parties? Should we show deaths from future heat waves? Decreases in the global wheat harvest? Something else? What do negotiators pay attention to?

She didn't pause to think for long. "We pay attention to how far we can go before our president or our prime minister loses the next election. That's pretty much it."

That exchange was a turn in the road for me, though I didn't know it at the time. It planted the seeds of two questions: Could we really solve climate change the way it was being framed if political leaders saw it mostly as a cost or sacrifice that their constituents would only stand for so much of? And if not, could we make more progress by widening the lens with which we looked at climate change, to include costs *and* benefits?

My search for answers led me forward to new research, new projects, and new partners. It also led me back to some of my own training and intellectual roots. And it led me outward, to other communities of practitioners, thinkers, and change-makers. Each of those threads—forward, back, and outward—has shaped the thinking in this book.

Let's start with forward. Not long after that climate negotiator set me thinking, I asked my colleague, researcher Diana Wright, this question: What else would be different in a world that had successfully addressed climate change? Diana brought me back a report that stunned me. After surveying studies in health, agriculture, water, and jobs, she estimated that the savings from being free of fossil fuels would balance out the costs required to reach that low carbon goal. That understanding is more commonplace today (though by no means universal). But at that time it shocked me so much that I asked Diana if she made a math mistake! Her report flipped my view of the climate change problem. I began to see that those leaders, afraid climate action would cost them politically, could truthfully promise their constituents cleaner air, better jobs, improved health, more energy security, and more resilience to natural disasters.

While Diana was conducting her research, I was also learning from my friend Angela Park, an expert in equity and justice in mission-driven organizations and movements. Her report, *Everybody's Movement*, was published the same year as the UN climate summit where I began to wonder about the possibility of widening the lens with which we looked at climate change.[12] Based on in-depth interviews with environmental justice leaders, the report made the case that the climate movement over-relied on science and policy goals instead of organizing, movement building, and connecting climate change to other struggles. As a result, society saw climate as an ecological issue, lacking immediate relevance to daily concerns, like family, health, community, or economic well-being. *Everybody's Movement* painted a picture of what might be possible if the climate movement was broadened, with openness to leadership from different segments of society than the "usual suspects." Angela wrote, "We must create new partnerships and a new framework, connecting seemingly disparate issues and addressing the systemic inequities and chronic dilemmas facing communities, people, and ecosystems across the planet."[13] Angela joined our project as a consultant and interviewed environmental justice leaders about how they saw climate change intersecting with other issues. In our conversations Angela was a strong advocate for rethinking climate change in ways that tied health, safety, economic opportunity, and equity together in a more integrated package.

While my thinking had been sparked at the level of UN climate talks, Angela was articulating similar conclusions from a very different starting point: for people in marginalized communities, climate change, inequities, health, and well-being were interconnected, not separate. Many fruitful conversations with Angela, dating back to that time and continuing over the years, have sharpened my thinking about multisolving, particularly when it comes to equity and environmental justice.

Spurred by these conversations, I started up a new stream of work at Climate Interactive, the organization that had brought me into the UN climate conferences in the first place. We began to document the multiple benefits of climate action and to profile "bright spots" of multisolving where people acted together to address climate change and solve other problems at the same time. Those bright spots showed that the potential gains in health, equity, or jobs weren't guaranteed. They had to be designed for. They took collaboration. There was an art to it. Wanting to better understand that art led to projects on green infrastructure, health, climate, and equity in Milwaukee.

During the work in Milwaukee, we didn't have a word for multisolving. We were using the term *co-benefits*, but it didn't fit well. If all the benefits mattered, which one was "main" and which was "co-"?

Angela Park was seeing something similar in her interviews with leaders who were addressing climate change in ways that helped meet other needs too. She described that way of working as akin to multitasking but without the negative connotations of being spread too thin or being distracted. Rather than attempting multiple tasks simultaneously, Angela pointed out in one conversation, we need approaches that solve lots of problems at once. Angela's observations reminded me of one of my favorite Wendell Berry essays, "Solving for Pattern," and we began using the term *multisolving* for what we were observing in Milwaukee and beyond.

After Milwaukee, our focus shifted to a collaboration in Atlanta, where I met another key partner who also influenced how I think about multisolving: Nathaniel Smith, founder and Chief Equity Officer at the Partnership for Southern Equity. Nathaniel's deep knowledge of equity, especially racial equity in the United States, influenced my thinking a lot, and I've done my best to acknowledge his influence in relevant places throughout the book. Tina Anderson Smith, who led the evaluation of our work in Atlanta, first exposed me to ideas like coherence and self-similarity, and you'll read about those ideas in the second half of the book.

My early training is the source of the systems framing of this book. Much of that framing I attribute to my friend, mentor, and boss Donella Meadows. She's known both for her contribution to the *Limits to Growth* project and for her book *Thinking in Systems*. From Donella I learned how to apply a systems lens to complex problems. She convinced me of the power of vision and the importance of questioning our mental maps and giving up our quest for control, all themes of this book.

In her essay "Dancing with Systems," Donella wrote:

> Systems thinking leads to another conclusion—however, waiting, shining, obvious as soon as we stop being blinded by the illusion of control. It says that there is plenty to do, of a different sort of "doing." The future can't be predicted, but it can be envisioned and brought lovingly into being. Systems can't be controlled, but they can be designed and redesigned. We can't surge forward with certainty into a world of no surprises, but we can expect surprises and learn from them and even profit from them. We can't impose our will upon a system. We can listen to what the system tells us, and discover how its properties and our values can work together to bring forth something much better than could ever be produced by our will alone.[14]

To the extent you find the spirit of working with systems—rather than trying to control them—throughout this book, you have the influence of Donella Meadows to thank.

While there are other books about systems, my approach is a little bit different. For this you have the many people to whom I've taught systems thinking to thank, especially the Donella Meadows Leadership Fellows. From these global sustainability leaders, I learned that you don't always need a diagram or a graph to impart important systems ideas. In that spirit, in the chapters that follow we'll approach systems through stories, metaphors, and even poems, more than through diagrams and charts.

Another thread I pulled into multisolving came from another mentor, Joanna Macy. Joanna is the developer of what she calls the Work That Reconnects, tools and practices for working with the strong emotions that the converging crises of our times can provoke.[15] One of Joanna's teachings is that there are many types of actions needed at this time. She says we need creative experiments (like eco-villages and resilient urban design), new worldviews, and holding actions that prevent harm. That's a way of describing multisolving: stopping the harm and creating the new all at once. Joanna's background as a scholar of both Buddhism and systems theory also shaped some of my thinking about the partnership worldview that is core to multisolving.

In addition to threads from my past and my own learnings (with partners) along the way, the ideas in this book have been influenced by many others also searching for ways to address problems in an integrated fashion. From public health experts measuring the lives saved by clean energy to labor researchers investigating the potential for green jobs, there's a growing body of scholarly research that has shaped my thinking about multisolving.

My thinking has also been influenced by the work of environmental justice leaders and the communities to whom they are dedicated. Health, justice, equity, energy, water, jobs, and climate all intersect in communities. The environmental justice movement has shined a spotlight on these interconnections for decades, dating back at least to the formulation of the Principles of Environmental Justice formulated by the first National People of Color Environmental Leadership Summit in 1991.[16] With their emphasis on solidarity, bottom-up leadership, and just relationships among collaborators, these principles of environmental justice offer a transformative approach to addressing environmental and social challenges. There's also the work of scholars like Dr. Robert Bullard, who is often called the father of environmental justice. His books focus on environmental health and racial justice and the impacts of events like Hurricane Katrina.[17]

My understanding of multisolving hasn't emerged solely from teachers and text, or from partnerships and practical experiments either. I also saw the power of multisolving firsthand (though I didn't have a word for it yet) in 2011, the year Tropical Storm Irene battered New England, including the region around my home in Vermont.

There I was, a climate researcher watching (and trying to assist in) the recovery from a disaster likely made more severe by climate change. What I noticed in the early days of the disaster was that it was food pantries, farmers markets, locally owned businesses, and churches that most helped my flooded-out neighbors. Those entities organized volunteers, collected and distributed supplies, fed people, charged phones, and provided drinkable water. Watching and participating in that recovery helped me see how the same dollar I donated to our local food pantry might contribute to both community well-being in normal times and resilience in dangerous times. That's the essence of multisolving, and it was made visible across my state during the Irene recovery.

A lot of threads wove together to shape the ideas of this book, and maybe that's appropriate. Maybe a book about solving problems by bundling them together can only be informed by a bundling a lot of different perspectives and schools of thought. Maybe a weblike approach to solving problems requires a weblike collection of teachers, mentors, and colleagues.

My hope is that such a book will be of interest to a web of readers as well. I came to multisolving via a focus on climate change, but you could just as easily pick up this book because you want to see more progress in health, equity, biodiversity, or economic vitality. In that sense, this book is for people who are alarmed about at least one problem and want to do something about it. If something in your world is keeping you awake at night, that's your entry point. I hope what you find here will give you some new ideas to try, either on your own or with others.

As we will see in later chapters, multisolving can be employed at a scale as small as a single neighborhood or hospital. It can also be employed region-ally, nationally, and at every scale in between. And no matter what scale interests you most, you can learn about the principles of multisolving from studying how it works at other scales too. The same goes for what sector you focus on. Multisolving can involve the collaboration of governments, nonprofits, businesses, and individuals. There's room for everyone, and I've aimed in my choice of examples throughout the book to showcase that diversity.

Is this book for everyone, then? Regardless of what problem or crisis motivates you, what scale that problem manifests at, and what sector you come from? Well, no, probably not. There's one more ingredient that will make this book worth your time: frustration with other approaches. If your efforts to address one or more of our converging crises (at whatever scale, from whatever sector) seem to fall short of what's possible, I hope you will find new ideas here. If you can see possible solutions but don't have enough allies (or dollars or shovels) to make them happen, I hope this book might help you see other routes to success. And if you just feel stuck or alarmed by the complexities and interactions of multiple converging problems, I hope that some of the tools in the book will help you make more sense of it all.

Many of you will read this book alone, but I hope that many of you will read it together with others. You'll find questions for reflection at the end of each chapter that could be fodder for your journal or your morning walk but also for a discussion group in your classroom, your community, or your organization. I hope some of you will read this book at the outset of collaborations, to have a common language and a library of solutions that others have used to overcome some of the obstacles you might find yourself facing along the way. Especially if you meet together with others to read this book, I hope the process itself will help you develop the friendships and connections that make multisolving possible.

Whether reading alone or in a group, you can also find more resources to support your journey through the book at https://www.multisolving.org/multisolving-the-book/.

A funder once asked my colleagues and me to conduct a global scan for multisolving. They were interested in examples from outside the United States that could be replicated here. We found many examples but also drew a conclusion I think they found surprising. We recommended that they *not* try to replicate the specific projects we had found. Each project fit perfectly to the specific geography and culture and needs of its place. We didn't see how those projects could be replicated elsewhere. Instead, we said the foundation should try to foster the *attitudes and approaches* of those projects. That's part of the beauty of multisolving. Every example has something to teach, even if the specifics of your situation are very different from mine.

Multisolving isn't a specific set of procedures one must follow. You won't find ten simple steps or cookie-cutter protocols in this book. What you will find are the best descriptions I could capture of those attitudes and approaches, the true replicable units of multisolving. As you read the following chapters, I hope you will listen for that vital underlying spirit.

Web World

Spiders worked all night
so that when I walked
   through the pasture
it was web filled,
like the world.
It was only the clinging dew drops
that made the webs
strung between meadow rushes
   and milkweed stems
visible at all.
This is how webs are: to see,
   look for sparkle.
There is a web strung by a
   butterfly between this
   milkweed and Mexico.

There is a water web that connects a
   dew drop to an upwelling ocean.
And
of course,
there is a web that connects
   you to me.
Moving through a web
   world could be a joy.
Or, it could hurt; it could desecrate.
Or—
you know this,
you can trust this—
it could heal.
It's not the web that tilts the balance.
It's how you move within it.

CHAPTER ONE

Multisolving: Promises and Obstacles

Some years ago, my colleagues and I began studying interventions that solve multiple problems at the same time. We found them everywhere, at both neighborhood and national scales and in every country we looked. We found them across sectors: in urban planning, health, agriculture, forestry, energy, transportation, and disaster management.

On the surface, these projects were all quite different, and the people undertaking them certainly didn't think of themselves as using any special methodology. Still, we found that the projects had much in common with each other.

It would be helpful, we thought, to find a word that would categorize these diverse projects. It seemed to us that they had much to learn from each other and much to teach the world. We couldn't find any word in use that quite captured these approaches and their potential impacts. So, we began to use the word *multisolving*.[1]

We defined multisolving as using one investment of time, money, or energy to address multiple problems. Once you start looking, examples of multisolving are easy to find.

For example, when a city greens by increasing its number of parks, gardens, and trees, it addresses multiple problems.[2] It improves the sense of well-being of residents. It reduces energy use for cooling and reduces the urban heat island effect, a burden of warming climates that disproportionately falls on marginalized communities. It can help reduce urban flooding by absorbing and slowing the flow of stormwater. That's multisolving: the single solution of greening the city solves multiple problems.

Multisolving can also address problems that play out on different time scales. For instance, reducing burning of fossil fuels addresses the health impacts of air pollution in the near term and protects the climate for the long term.[3]

Importantly, multisolving can address symptoms and root causes at the same time. That's because so many of our current crises are the result of a worldview of disconnection and separation. Multisolving happens best when people adopt a worldview of interconnection. So, a community that comes together to address flooding, energy bills, and the impact of climate change on its most marginalized community members is addressing symptoms: flooding, economic hardship from energy costs, and high heat days. But by working together to benefit many issues at once, the community is also forging new relationships and working together in new, more connected ways. They are solving problems from a different frame of reference than the frame that created those problems.

Multisolvers around the world are showing us what might be possible if more of our efforts to address one problem could address several all at the same time. They're demonstrating another way to steer systems in new directions, toward justice, health, equity, and sustainability.

In tumultuous times, we know we need to navigate short-term crises while also steering systems in a different direction for the long term. Multisolving is one powerful approach for that. And it's available to us all.

Although our team began using the word *multisolving* in 2015, the idea of solving several problems with one effort is, of course, nothing new. I grew up hearing the adage "kill two birds with one stone." While a little bit gruesome, the saying does convey the essence of multisolving. My grandmother, who raised her family through the Great Depression and the Second World War, was a great multisolver, out of necessity. Quilts kept the beds warm, but she could also use them to create a makeshift room around the wood stove to create a steamy spot and help her croupy baby breathe easier. The church suppers she helped organized fed people, raised money for the church, and helped maintain a sense of community.

The small farmers I know are master multisolvers. Scraps from the table become food for the pigs, solving a waste disposal problem and a cost-of-feeding-pigs problem. In his essay "Solving for Pattern," Wendell Berry described the work of managing a successful farm as optimizing for many things rather than prioritizing just one: "The farmer has put plants and animals into a relationship of mutual dependence, and must perforce be concerned for balance or symmetry, a reciprocating connection in the pattern of the farm that is biological, not industrial, and that involves solutions to problems of fertility, soil husbandry, economics, sanitation—the whole complex of problems whose proper solutions add up to health: the health of the soil, of plants and animals, of farm and farmer, of farm family and farm community, all involved in the same interested, interlocking pattern—or pattern of patterns."[4]

Permaculture practitioners refer to the principle of "stacking functions," getting multiple types of use out of each element of a permaculture design, such as a plant that produces both food and fiber, or shade and food, or even all three. Permaculturist and author Toby Hemenway described this idea in a 2014 interview: "Nature is marvelous at what we call 'stacking functions,' where if you have a conventional landscape designer, they may choose a tree for shade or fruit or, you know, a single function. And, if you look at what a tree is doing in any natural system, it is producing fruit, it is producing shade, the leaf litter is building soil, the roots are breaking up heavy soil, it is harvesting rain and channeling it somewhere. It is habitat for a zillion different kinds of creatures. And, that is the kind of thinking that permaculture recommends."[5]

Though not so named, the concept of multisolving is also a part of many Indigenous knowledge systems around the globe. In his book *Native Science*, Tewa author and professor Gregory Cajete describes the traditional planting techniques of some Indigenous peoples in North America: "When soil was planted with the traditional 'three sisters' of corn, beans, and squash, with other native plants such as marigolds nearby, corn provided shade for the delicate beans and a stalk on which the vines of squash and beans could grow. The squash provide extensive ground cover, reducing weed habitat and weeding, and simultaneously shielding the soil from rain erosion while capturing a maximum of available rainfall. Beans fix nitrogen in the soil through their roots, and so improve fertility."[6] This way of thinking may be common and ancient, but before our team began using the word, I couldn't find a simple term that seemed to capture its essence.

At the time (and still today) one way to talk about projects that meet many goals is to refer to "co-benefits." But co-benefits imply a "main" benefit. Often that main benefit is something global or something off in the distant future, like climate protection. Is climate protection the main benefit of closing a coal-fired power plant? Maybe, but if you are a parent to a child with asthma, the benefit of cleaner air might feel central, not "co-."

Co-benefit language, perhaps unintentionally, implies a hierarchy of benefits, with the long-term and global often seeming more important than the short-term and local. The point of multisolving is that it doesn't matter which benefit is core for you and which is core for me. If we can collaborate to accomplish both benefits, then that's a win for both of us. One reason I like the word *multisolving* is that it frames all the benefits of a project as important. The word is a reminder that all the constituencies being served are important and puts all goals on equal footing. The word reminds us that we are all in this together.

Another advantage of the word *multisolving* is that it can be a verb. Multisolving is something you *do*, an action you take, as opposed to a result (like a co-benefit). The word itself is a good reminder that multisolving is a way of working one can cultivate and improve at over a lifetime.

Systems

Multisolving is an approach that is designed for complex systems. A system is an interconnected set of elements that function together. Your body is a system, of course. Within your body, your liver is also a system, and so is a liver cell. Book clubs are systems, as are neighborhoods, political parties, grassy meadows, and tide pools. Systems are complex. Often, they are highly interconnected. Their behaviors change as the connections within and between them change.

Your family is a system that influences both political and ecological systems. And of course, those political and ecological systems turn right around and influence your family.

Each of us is a system nested within multiple systems, subject to a vast array of interconnections. In fact, the cascading and converging of problems described in the Introduction stems from these interconnections.

Interconnections are why one problem can make another worse. A single crisis, via cascades and ripples of cause and effect, triggers others. Reinforcing feedback can amplify problems, which can then grow so large that they push systems across unexpected thresholds, giving rise to entirely new problems. The complexity of systems gives us plenty of good reasons to be concerned about problems interacting and amplifying each other.

But the highly interconnected nature of systems isn't always bad news. It also means that, sometimes, a single intervention can set off ripples of connected solutions. Some problems become easier to solve when you tackle them together. One well-crafted intervention can solve many problems at the same time.

Interconnections between systems are neither good nor bad. They just *are*, a feature of the world. Multisolving acknowledges and works with these interconnections when more common approaches often ignore them. When money, resources, time, and attention are in short supply, this is great news. We can use a system's interconnection in positive ways. We can focus on interventions that ripple outward, solving many problems at the same time.

Here's an example. A community organizing to shut down a coal-fired power plant will, if successful, solve several problems. For one, they will improve the air quality of the neighborhood. Less air pollution—which contributes to many illnesses, including respiratory disease, heart disease, and premature birth—means the health of people in the neighborhood will improve. Less coal being burned also means less of the greenhouse gas CO_2 will be generated, so the project will protect the climate. One action will have improved air quality and people's health in the near term and helped protect the global climate in the long term.

One of the beauties of multisolving is that anyone can do it.

Individuals can and do multisolve, in both their personal and professional lives. When a young father fills his shopping cart with sustainable and fair-trade products even though they cost a bit more, he is multisolving, his family budget helping to protect health, livelihoods, and ecosystems near and far from home. When a retiree offers landscaping classes through her church that teach people how to install rain barrels, raised garden beds, and plants to support native pollinators, she helps build community connections, conserve water, feed people, and protect biodiversity.

The small farmer who produces healthy food while restoring the land that she tends and protecting habitat is multisolving. The teacher who leads his class to install a solar panel on the school roof and batteries in the basement is multisolving. His project reduces greenhouse gas pollution, prepares students for roles in the clean energy economy, and provides the neighborhood around the school with a source of resilience in the face of outages in the electricity grid. The energy minister of an island nation can multisolve when she directs investment into clean electricity and closes down a fossil fuel–burning power plant.

The possibilities for multisolving become even more far reaching when the process is approached together with others.

Policymakers can work across the typical silos of government to multisolve. Housing budgets can be deployed to house people, reduce energy use, lessen the need for cars and highways, *and* foster economic equity.

A country's energy policy can conserve energy *and* improve national security. Global climate policy can serve international development goals *and* help to reduce poverty.

Collections of businesspeople, citizens, nonprofits, and government entities can collaborate so that new infrastructure provides economic opportunity for communities who have been historically marginalized. They can make sure the infrastructure helps meet environmental and health goals while also supporting local small businesses.

Reasons to Multisolve

We're more used to breaking problems down into their smallest components rather than looking for ways to bundle multiple problems together. Because multisolving approaches are countercultural in this way, they can feel unnatural to attempt. And as we'll see in the following section, there are obstacles to multisolving that take effort and planning to overcome. Before looking at the obstacles though, let's explore some reasons why multisolving is worth the added effort.

Make efficient use of investments of time and money. We need investments in new infrastructures and technologies. We need investments to repair past harms. We need investments to adapt to change already in the pipeline that we can't stop. The deeper we move into overshoot of sustainable limits, the more these needs are likely to grow.

But the time, money, and attention we have available to address these needs are limited. One of the biggest challenges of our time is to solve as many problems as we can with the resources we have.

Faced with this reality, multisolving is basic common sense. In our daily lives we are always looking for such synergies. Can we paint the community building and strengthen the community fabric at the same time? Can we serve local vegetables in the school cafeteria, supporting both our kids' health and our neighbors' farms? Multisolving is the original "two for the price of one" bargain.

The wisdom of multisolving was demonstrated by some city, state, and national leaders around the globe in response to the COVID-19 pandemic, when stimulus packages that emerged to address the pandemic's economic slowdown were invested in ways that accomplished multiple goals.

In Nigeria, some economic recovery spending was directed toward building solar electricity microgrids to replace polluting and health-harming fuels like kerosene and diesel.[7] Communities that couldn't access the central grid previously now have clean local electricity for lighting, refrigeration, and businesses.

In Medellín, Colombia, stimulus spending increased the number of bike paths and public transit routes.[8] Such projects create jobs, reduce greenhouse gas emissions from transportation, and provide access to mobility for city residents.

Weatherizing the homes of low-income residents was a provision in many stimulus packages. Weatherizing contributes to economic recovery by providing good jobs, contributes to climate change protection by improving energy efficiency, and—because low-income residents pay the largest share of their paychecks on energy—reduces economic inequity.

The pandemic created unprecedented need for assistance and support, and governments faced tough choices about how to target aid. These multisolving stimulus packages maximized the impact of every dollar.

Build power and overcome resistance to change. While the costs of some actions in systems are upfront, their benefits may not be apparent until some unspecified time in the future; for example, we need to build a clean energy infrastructure today to avoid worse climate outcomes decades from now. This can lead to political resistance, when elected officials who are concerned about reelection delay or avoid strong action on climate because voters perceive a difficult cost in the present for a benefit they may doubt or never see.

Resistance to change can also occur when a cost is felt locally, while the benefit happens in a far-off place. Actions to build a clean energy system in my state show up as an increase to my state tax bill, even though reduced emissions in my state benefit not just my state but the whole world.

Multisolving projects, which tend to combine local benefits with global ones, can help overcome both of these sources of resistance to change. When my state enacts an energy efficiency program, we contribute to long-term global climate protection. We also harvest some local immediate benefits, including good jobs, cleaner air, financial savings, and less reliance on distant energy sources.

On a smaller scale, a company might invest in an upgrade of the physical plant to allow more natural light and better insulation and windows. Looked at narrowly, that would register as a cost to the capital budget. But studies have shown comfortable spaces infused with natural light make for happier and more productive employees.[9] Within a company, the advocates for this sort of project would be wise to share the full mix of benefits when they seek approval. They could explain that the costs on the capital side of the budget would be offset by the gains on the operating budget.

Design investments to meet multiple goals. Help stakeholders understand the full picture. Leaders who do both of these things can help ease resistance to change.

Use synergy to prepare for the future while tending to the present. When it comes to climate change and other crises that we cannot fully prevent, we know that our communities, economies, and societies must adapt. Investments in adaptation can be costly, from building giant infrastructure projects to hardening coastlines, moving buildings and equipment to higher ground, redesigning highways and railroads to withstand stronger storms, and preparing cities for more extreme high heat days. If these investments are required to keep people safe in times of extreme impacts, wouldn't it be wonderful if we could harvest other benefits from them as well? That's another type of multisolving.

Can a strategy to protect a coastal area from sea level rise also protect habitat and biodiversity and fishing economies? Since a stretch of road needs to be rebuilt to accommodate increased stormwater flows, can a cycling lane be installed as well? When new housing is built outside of the floodplain, can it be energy efficient, powered by the sun, and clustered in a cooperative neighborhood with dedicated green and gardening space? With planning, intention, and cooperation across silos, it is possible to brace for future risks while also improving people's quality of life right now.

Embed justice. To respond to climate change, we will need to build new infrastructures in the span of mere decades. Every community will need access to zero-carbon electricity. We will need to build millions of new structures to net-zero standards. We will need to upgrade millions more to those same standards. How we use land will need to change at much the same pace. Agriculture will need to shift its practices, as will forestry. And so on.

Entirely new infrastructure sounds like a big job, but that's only part of it, isn't it? On top of that is making sure everyone is able to meet their basic needs. When almost 10% of the world's people are currently living in extreme poverty as defined by the World Bank, and 44% on less than $5.50 per day, that is a huge challenge.[10] The changes we need to implement must take care of more people than ever *and* relieve pressure on the Earth *and* withstand the effects of climate change we are unable to prevent.

Multisolving will be essential to doing all of this well. Imagine what might be possible if all the massive changes the world is beginning to embark on practiced multisolving. If every new technology could be shaped to accomplish more than one objective, and with influence from more than one kind of expert. If every bit of new infrastructure to meet the challenges of climate change and habitat loss could be designed with input from both climate change adaptation experts and the local people who would live with its impact, as well as designed to provide economic opportunities in ways that would reduce inequity.

And imagine how wrong it could all go without a multisolving approach. Imagine trillions of dollars invested in low-carbon infrastructure only for it to succumb to future climate change impacts. Imagine new infrastructure that exacerbates rather than heals inequalities.

Both futures are possible. Our commitment or lack thereof to multisolving will make the difference.

Avoid misguided solutions that create harm. The scale of the transformation needed to address crises like biodiversity loss and climate change is staggering.

For instance, the International Energy Agency says that to mitigate climate change we will need to electrify much of the infrastructure currently powered by coal, oil, and gas. That means batteries, lots of them, which will require mining an estimated forty-two times more minerals like lithium by 2040. Lithium is currently extracted at great environmental and human cost in countries like Chile. Without a multisolving approach, a climate "solution" could worsen the well-being of communities near lithium sources. A multisolving framework broadens both the questions and the design challenges. How can we meet our energy needs without fossil fuels *and* without harming the places that produce essential components for renewable energy? Holding both of those objectives at once leads to new ideas, from careful recycling of batteries, to reducing energy demand by changing patterns of consumption and urban design, to supporting democracy and local control in the places where raw materials are extracted.

The biodiversity crisis poses similar challenges. Land conservation is an essential tool for the protection of species. A multisolving approach to land conservation would place leadership of such efforts in local hands. This includes returning land to Indigenous peoples who, while currently stewarding only 20% of Earth's land, protect 80% of global biodiversity.[11] It would include livelihood and economic well-being as another design goal alongside conservation. Without this perspective, a singularly focused conservation program risks creating displacement and poverty. And because local communities and Indigenous groups hold key knowledge about managing lands, the conservation approach itself needs their leadership and input.

Act from a worldview of connection and interdependence. The primary purpose of multisolving is to help people address problems more effectively. But multisolving does something else.

So many of the tangled and converging crises that we face have their roots in a worldview of disconnection and separation, a style of interaction some have called power-over (whereby one group tries to thrive by dominating another). Instead of participating in the closed loops of ecology, the global economy as it is currently configured extracts resources and deposits wastes, giving rise to resource crises, biodiversity crises, and pollution crises. Instead of sharing burdens and benefits equitably, our economic and governance systems too often prioritize the well-being of one group over others. This pattern creates other crises, including gender inequity, racial inequity, religious and ethnic inequities, and sexual orientation inequities.

Multisolving requires people to work together across the parts of a system. They have to replace disconnection with connection and power-over with power-with. They have to heal the fractures created by dividing the world up into disciplines, departments, and jurisdictions. They have to reach across the boundaries of different communities or different cultures. They have to grapple with artificial constructions of "us" and "them." They have to listen, connect, and find the points where multiple goals rise together.

This means that even while multisolving addresses immediate, tangible problems, it does so in a way that is distinct from the habits of thought and action that created those problems in the first place. In this way, multisolving is an antidote to the predominant, anti-systemic way of looking at the world.

Long after a particular multisolving project ends, the experience of working together as a whole system stays with people. It offers a template for future work and another way of looking at problems.

Boost adaptability. Because multisolving projects happen when people work together across the typical divisions in our world—departments, jurisdictions, disciplines, and more—multisolving projects leave behind a system that is connected in new ways. Long after the orchard is planted or the jobs program is funded, the people who pulled all of that off remain connected to one another as friends, colleagues, and trusted partners. That is a source of resilience. When conditions change and new problems or new opportunities arise, the networks created in the process of multisolving are a source of adaptive capacity. People know who to call for help. Leaders trust each other enough to move quickly and collaborate well.

Live your values. "The well-being of everyone in this community matters to me."

"I don't want my long-term solutions to come at the expense of someone else's health or prosperity."

"Future generations have a stake in what we decide today."

"I work in the arts (or I write cookbooks or design apps or teach preschool or . . .), and I want that work to contribute solutions to global problems and to protect future generations."

If sentiments like these fit in in your personal code of ethics, then you have one more reason to multisolve. Your values call you to work in a way that includes concerns and interests outside of your own narrow sphere. Your values will lead you to consult others, to look for unanticipated impacts, to draw wide boundaries in your mental map of systems. That, of course, is just the stance for multisolving.

Can the new wing of the museum also be a showcase for low-carbon design? How can all children in the community benefit from the exhibits they will pass through?

Start with the problem that keeps you awake at night and ask what other problems, keeping other people awake at night, could be solved by addressing yours. In doing so, you will be living out your values. You'll also be well on the way to multisolving.

Obstacles to Multisolving

Although my colleagues and I found multisolving projects everywhere that we looked, on every continent and at every scale from small towns to nations to regions, it remains rare. If that sounds paradoxical, just look at your own world. Do most people work in organizations that encourage them to explore widely, collaborate, and work across different domains? Do most governments have mechanisms that allow employees to work across departments? Do most universities make it easy to be interdisciplinary? In my experience, the answer to all these questions is a resounding no!

Although multisolving seems to be possible everywhere, it isn't nearly as widespread as the need for it, or the opportunity. Why is that? In this section we'll look at some of the reasons that help explain the relative rarity of multisolving.

Disciplines. Universities have departments and disciplines. Academics publish and converse within their separate disciplines. Each field evolves its own specialized language and a knowledge base that can be daunting to outsiders. The people who study the impacts of air pollution on the human body have degrees in medicine and public health. The people who study the impacts of greenhouse gas pollution have degrees in fields such as climatology or ecology. Then there are the engineers who design the energy systems that combust fossil fuels to produce both air pollution and greenhouse gas pollution, and the policymakers who set the incentives within which the engineers must work. It's the rare individual in any of these fields who was educated in a way that explored all these connections. And after graduation, it's the rare professional whose job description allows them to step back to view the full picture. There may be pathways to the future where all these interests rise together, but whose job is it to find them?

Restoring ecosystems can help protect biodiversity and help build resilience to climate change impacts. That work can also provide good jobs for people who need them. But when does a career in youth job training intersect with one in biodiversity? Until very recently, the odds of such a connection were slim, and even today, with growing interest in ideas like a youth climate corps, leaders are feeling their way into the intersections of very different fields.

Budget silos. Multisolving provides a net benefit for the whole system. But sometimes costs burden only one subsystem while, at least initially, benefits accumulate in another. From the vantage point of the first subsystem, the entire project may not seem worth it in the early days.

Consider a country contemplating replacing fossil fuels with clean energy. Where would the costs fall? Probably on the transportation and energy ministries. Where would the benefits accrue? Well, there'd be fewer hospitalizations from air pollution-related illnesses, representing a savings to the health ministry. The benefits to the health system might match or even outweigh the costs to the energy and transportation systems—a report from the World Health Organization found that, worldwide, the costs of meeting climate goals would be offset by the savings to the health sector.[12] But in most countries, decisions about transportation and energy are not made in consultation with the health ministry. From the point of view of the energy minister or the transportation minister, transitioning to clean energy would mean more cost than benefit. The health savings probably won't even show up on the energy minister's annual budget report.

Smaller systems face the same difficulty. Trees, gardens, and other green spaces can be more expensive for public works departments to maintain than pavement. And while greening a city might produce health benefits, the public works department doesn't typically get "credit" for them.

Sometimes the challenges of paying for multisolving depend on a time dimension that influences budgeting. For instance, investments in the energy efficiency of a business's physical plant are a line item on its capital budget. Those investments will produce a savings on the operating budget. Perhaps, over time, they will even deliver a net savings for the business overall. But if one decision-making unit feels the cost and another gets the benefit, a good idea might never get off the ground without a way to look at the situation with a broader perspective.

In a similar vein, the health system savings from reduced air pollution from reduced use of fossil fuels is a future savings. But it may take some innovative financial instruments to tie that future benefit to the current expense of building more renewable energy.

Jurisdictions. Sometimes potential solutions to problems are to be found outside the jurisdiction of the people charged with managing a problem. Urban flooding can be reduced by protecting upstream wetlands, but those wetlands can sit in a different county or even a different state than the leaders trying to cope with flooding. A fisheries manager might be trying to manage nitrogen pollution that is creating zones of low oxygen. But the source of that nitrogen might be in the practices of farmers hundreds of miles upstream.

Waterways, foodsheds, transportation networks, migratory routes of animals, air currents that disperse pollutants—none of these conform to the boundaries of political jurisdictions. Making headway on such issues asks us to find ways to work together across boundaries.

Time pressures. The same features that make multisolving so powerful can make it feel slow, messy, and inefficient. After all, a meeting with one or two decision-makers is likely to reach a conclusion more quickly than one with a half dozen different organizations.

The more interests that you bring into a problem-solving space, the more concerns and perspectives you must address. Each of these considerations will take time for all parties to understand. People might need to learn new technical language from other fields. They'll have to find their way around obstacles they never encountered before. There could be arguments or conflict.

Leaders within systems often feel pressure to bring projects to quick conclusions. This incentive prevents some from even trying to multisolve. A colleague once reached out to me excited about the potential for multisolving at the state level. She was a member of a task force involved in updating the state's energy and climate plan. She reached out to the task force chair, suggesting a multisolving approach.

The task force was already quite diverse. It included experts in health, jobs, transportation, and housing, in addition to climate and energy experts. To my colleague, it seemed to be a perfect setup for multisolving. But the idea was quickly squashed. The chair understood her job to be getting a blueprint on the governor's desk as quickly as possible. She wanted to focus narrowly on tons of greenhouse gases and not get distracted by complexities of health and jobs and equity.

That may have been the right call given the task force's tight timeline and the limited resources at the chair's disposal. The conditions set in the governor's office wouldn't have supported a robust multisolving process. Still, this story is a good example of why multisolving is not more common. Multisolving doesn't just happen. It requires time, space, facilitation, and a modest amount of funding to support its activities.

Often a crisis or an emergency highlights the need to multisolve. But emergency conditions aren't always the most conducive to experimenting, forging new relationships, or learning new things. It's hard to slow down and ask who else should be present in a discussion if it feels like something precious is under threat. In a diffuse organization or government, it might be hard to trust that you can afford to take the time to get to know colleagues and potential partners and see problems through one another's eyes.

When vulnerable people are facing imminent harm, an emergency response is needed. Not every moment is a moment for the relationship building and learning that help multisolving thrive.

But often there *is* time to widen the conversation and listen to one another and work together, or there could be. And given that so many of the crises we face were created by decision-makers working in silos, maybe monosolving isn't actually as efficient as it seems!

The standard for an "efficient" process is often "be fast and lean," rather than "get us to a good and durable result." That standard can be a barrier to multisolving.

Resistance to change. Multisolving requires people with different expertise to come together to implement solutions. A transportation planner might need to engage with experts in poverty, jobs, green space, parks, and health. He won't be an expert in those arenas. He will need to be a learner.

Curiosity and the ability to respect the expertise of others are attributes of effective multisolvers. For some, working across perspectives like this feels exciting. After all, there's so much to learn from other fields and other ways of looking at the world. But many of us were trained within institutions that reward specialization. Expertise is a source of respect and prestige. Sometimes it is a means toward organizational advancement, a higher salary, and more autonomy. Because of this, expertise can understandably become both an identity and a source of security. Not knowing, needing to ask questions, and being uncertain can make someone feel vulnerable in multisolving settings.

Multisolving definitely needs experts. Ideally, each participant in a multisolving effort brings state-of-the-art knowledge and practical experience. But in multisolving projects, knowledge is not useful unless it can be woven together with other knowledge. Can you be secure in what you know but comfortable with what you don't? For people used to identifying as experts for prestige and self-protection, that can be a real challenge.

If several people involved in a project are wedded to their identity as experts, it can be a source of resistance to change in the search for multisectoral solutions.

Pressures and incentives within organizations can be an obstacle to multisolving. The board of directors at a foundation may expect to be able to see that their donations have made possible a specific change in the world, as in "See that? We made that happen." A new wing of a museum, a new department at a university, a newly planted tree with a plaque on it.

A leader of a community organization may be trying to build that organization's reputation. He may seek a signature project. Then everyone, from the press to the mayor's office to the local funders, will know about his organization's brilliance.

When people working within systems expect to be able to point to their own signature of change and look askance at collaborations where everyone had a hand, it's hard to multisolve. Multisolving emerges from the connections made across many parts of a system. All participants deserve some credit; none deserve sole credit.

Vested interests within subsystems can be another source of resistance to change. Overall, a society might be much better off with less toxic pollution from plastics (or fossil fuels, or deforestation). But those whose careers and investment portfolios are tied to those subsystems will likely push back about multisolving. In later chapters we'll discuss some strategies for this pushback.

Inequity. So many of the systems we live and work within have been shaped by inequity. Whether it's the economic inequity between two neighborhoods in the same city, the racial inequity in the impacts of pollution, or the disproportionate impacts of climate change on the Global South, the people who come together to try to multisolve often must do so against a backdrop of inequity. Some organizations in a collaborative may be well resourced while others are barely scraping by. Some participants may have lots of formal training and positional power while others have lived experience but few "credentials." To truly work and learn together, these inequities will have to be faced, but it can require extra support to build the capacity for effective discussion about inequity and steps to address it.

The quest for control. Many change projects are designed to reduce uncertainty and control outcomes. This is understandable, of course. When people put time and money into addressing an important problem, they want to be sure to get results. Multisolvers want results too, after all. However, focus on predetermined outcomes can be taken so far as to make multisolving difficult.

A project funder may ask for highly detailed plans. What exact interventions will be used? How will they produce desired results? Can you spell out the steps involved?

On the surface these seem like quite reasonable questions, but as we'll see, multisolving is an emergent process. New ideas and possibilities are created by people coming together across silos. The resulting emergence is unpredictable. You can't fully know how a whole system will behave once its parts have been connected in new ways.

The first step of a multisolving process might be for people to get to know one another and explore the different problems with which they are struggling. So far, so good; that's a process one could explain to a funder or to those invited into a project.

But what will happen next? Well . . . it depends. It depends on the connections that are made. Who would like to work together? What new ideas bubble up when our different types of knowledge are pooled? Those are hard answers to predict in advance.

Later stages in a multisolving process might indeed involve specific projects. Two partners might plan a training or a policy intervention or a new piece of infrastructure. But it's unlikely anyone anticipated those priorities in the initial stage of allowing connections to emerge. If the process of kicking off a new project required such specificity, the good idea might never come to fruition.

Donors and project leaders aren't the only people who are susceptible to this challenge to multisolving. Potential project participants themselves, steeped in ways of working that don't make allowances for emergence, may find themselves confused or distressed. You'll find lots of ideas, when we talk about emergence in later chapters, to help support people's tolerance for the uncertainty of multisolving.

———~⚹~———

Any one of these obstacles is a barrier to multisolving. Taken together they might be enough to dampen anyone's enthusiasm. Single-issue problem-solving might sound easier. But is it really? If tangled and interconnected problems yield best to interconnected approaches, then monosolving might be a quick and efficient route to . . . failure. Like it or not, in these tumultuous times, we need to do our best to work with interconnection instead of looking the other way.

That's what the rest of this book is about: ideas, examples, tools, and resources for multisolving, so that you can address the issues that concern you most. For many of these barriers, the best antidote is success. Even small wins help others begin to understand what is possible.

So, to help increase your odds of success, in the chapters that follow we will explore the key elements of multisolving. We will start with systems: understanding them, working within them, and steering them. Multisolving is possible because of the interconnectedness of the world, and systems theory gives us a way of understanding, describing, and working with that interconnectedness. Then we'll explore how multisolving can be pursued, and the values, attitudes, and worldviews that best support it. We'll look at building networks and nurturing relationships, and I'll share from my experience how multisolving projects grow and evolve over time.

A chapter on equity will follow, both because inequity is a crisis that can be addressed by multisolving and because equity is a key value and practice for successful multisolving. Then, because whatever you put into practice from this book will likely play out amid increasingly destabilized earth/economy/human systems, we'll explore ways to prepare for shocks, exponential change, and other surprises. Finally, we'll put it all together and, if I've done my job well, you will realize that you already have much of what you need to multisolve.

That may sound like a lot, but we will get through it step by step. The place to start is with the elements of systems. So, onward to understanding systems. We will start with stocks, one of the most visible parts of systems.

? Questions for Reflection

- Pick one of the problems that worries you most, whether global, local, personal. List as many other problems as you can think of that could be improved by a careful solution to the problem you've identified. It's okay if you don't have the resources or the connections to implement the solution; allow yourself to dream a little and imagine possibilities.
- If you are reading this book as part of a group of people who live or work in community, compare your answers to the question above. Which problems showed up on multiple lists? Did your view enlarge at all by hearing others' answers?
- A system is an interconnected set of elements that function together. Name some of the systems that you participated in as you went about your day today.
- Look back at the reasons to multisolve mentioned in this chapter. Which of them speak to you the most and why?
- Are any of the obstacles to multisolving mentioned in this chapter familiar from your own life? Which ones show up most prominently? Which ones most impede the goals that you have? Have you found any effective solutions?

Turnover

I went back to our childhood pond
with its cool depths and ferny banks.
Remember how the mud
    would squish,
silky and warm, between our toes?
We'd catch green-jeweled bullfrogs
and hold them, slippery in
    our cupped hands,
until they'd squirm out of our fingers
back into the watery edges,
and the green sawtooth sedges.

I went back to the pond, and
    it was the same.
The same glassy stillness at dawn.
The same cool depths, and, at the far
    end, the same tumbling brook.
The same mud and sedges
    and mossy outflow.
As though no time had passed.

But of course. . . .

Not the same frogs.
Not the same water we
    floated in, on our backs,
    squinting up at the sky.

New water: rainwater, cloud water,
    spring water, brook water,
deep Earth water from
    nobody knows where.
Even while I stood there—muddy
    toes, shoes in my hands—
the brook rushed in.

Even while I stood there—
    breathing in, breathing out—
the water slid over mossy rocks
    toward the distant sea.
I stood still (or so it seemed)
    but the pond didn't.

The pond, our pond,
it turns out,
was never a thing.

And, my friend,
neither were we.

CHAPTER TWO

Stocks

It's hot in July in my little writing cabin, and Marie's Pond is just steps away. I jump in, float around, come back and write for a few hours, repeat.

Marie used to live in the farmhouse above the pond. She moved away a few years ago, but we all got used to saying, "I'm heading over to Marie's for a swim," or "Want to meet up for skating and a bonfire at Marie's?" I think we'll probably keep calling it Marie's Pond for a long time. In small towns, some names stick.

Marie's Pond is a manmade pond, dug out by a particular man, Will Curtis, in the 1950s, when the land was part of his dairy farm. Our best guess is that he scooped the pond out of a marshy wetland so that his cows would have a constant source of water, but nature has claimed Marie's Pond now. It's surrounded by alders, willows, and cattails and full of harrumphing bullfrogs and floating salamanders. Red-winged blackbirds nest in the alders, and, especially in the mornings and evenings, barn swallows swoop low over the water, catching bugs.

Springs are the main water source for the pond, welling up icy cold from the bottom; you can feel it on your legs as you swim over. The pond drains into Sugar Brook, the winding little stream that flows through this narrow valley.

That's Marie's Pond, but now bring your favorite pond to mind. I bet you have one. A crystal clear cold lake in the mountains? A watering hole in the grasslands? The place you learned to swim? A fountain in a park where toddlers splash on hot summer afternoons? A pond is any place where water accumulates, and ponds are also stocks—the first element of systems that we will explore.

Ponds (and bathtubs, tea kettles, and the Pacific Ocean) are stocks of water. A library is a stock of books, your pantry is a stock of food, a school contains a stock of children, and so on. In a systems view of the world, most of the tangibles you see are stocks. Little ponds and big ones, collections of material that move from pond to pond. If you can measure it or count it, odds are it is a stock.

Since you're familiar with ponds, you already have a good feeling for stocks. You know how stocks appear permanent and stable. After all, ponds tend to be there day after day, taking up more or less the same space, with the same outline. But the volume of Marie's Pond, like that of all ponds, ebbs and flows. Its boundary creeps up into the alders with the snowmelt and spring rains. In the heat of August, it recedes, leaving a little muddy zone where frogs and turtles bask in the sun.

The water in Marie's Pond is never the same water moment to moment, let alone year to year. This illustrates a really important point about stocks so I'll say it again: at any moment you can count up the contents of a stock, but the contents themselves are on the move, entering and leaving, sometimes in hundredths of seconds, sometimes over thousands of years.

Water flows into Marie's Pond from the rain and from the upwelling springs. It flows out via evaporation back to the sky and via an outflow to Sugar Brook. Stocks, including ponds, can seem permanent, but they are always changing, always turning over. Imagine a school. It's always full of children (at least during the day), but new kindergartners arrive each fall while new graduates leave each spring. And at night, during school vacations, and over summer holidays, the school is empty of children entirely. It is a stock that drains to zero. Far from being constant, permanent, or solid, stocks are temporary holding pens for material in flux.

Your pantry fills with groceries brought home from the store and empties when you cook and eat dinner. But the story of the pond that is your pantry doesn't begin or end at the grocery store. Prior to the grocery store, your pantry's contents were part of multiple stocks called warehouses. Before that they filled stocks including processing plants, silos, refrigerated trucks, fishing boats, and agricultural fields. And so on.

For a short period of time after you eat dinner, the food that was in your pantry is now in the stock called you, until it moves on again on into stocks of wastes and exhalations, some of which then become part of bacteria or soils or an ocean.

Not all the stocks that matter in systems are physical. Your bank account is a stock, one that may be growing, draining, or remaining at about the same level. The mutual trust between us is a stock. It fills when I act in a trustworthy way and drains if I lie or cheat. Knowledge is yet another intangible stock, as is goodwill or fear.

If you are wondering if some part of your world is a stock, here is a good test: Is the name for it a noun? Most stocks are filled with "stuff," things that, linguistically we tend to categorize as nouns. Cars, cows, books, happiness, trust, strength. If the variable you are thinking about is a verb, then it's likely not a stock. Words that convey motions or change—for example, growing, shrinking, moving—are likely not stocks.

Not all stocks are "good," and adding to a stock isn't always desirable. The filling of some stocks can create problems or make existing problems worse. For example, carbon dioxide pollution in the atmosphere is a stock, one that has been filling up since the beginning of the Industrial Revolution and one that puts our collective future in grave danger. The amount of endocrine-disrupting chemicals that have accumulated in my body is a stock that could cause health problems. People living in poverty or in hunger make up a stock, one that society would like to be low or zero. Power is a stock, including power that can be misused for harmful purposes.

With some stocks, everything is just fine when at a certain level of "fullness," but beyond that level (or below it) problems appear. The stock of potato beetles in my garden is manageable up to a certain number of bugs per plant, but if my efforts to pick the beetles off plants can't keep up with the growth of this stock, the beetles can strip the plants of their leaves in a single day.

Sometimes the most important thing to do to improve a system's health is to drain a stock. So, as you grow in awareness of the stocks, look not only for those that are too low and must somehow be filled but also for those that are causing harm and must be drained.

There's No Away

The water in Marie's Pond is connected to all water on Earth. It is only a way station in a global cycle that moves water from pond to stream to ocean to clouds. From the clouds water falls on watersheds, including, once again, ponds.

The bullfrogs, dragonflies, deer, and raccoons that drink the water are connected to the pond too. Water that was once pond becomes a stock of water in a deer or raccoon.

The water cycle stretches back to the early days of the Earth and will continue forward millions of years into the future, each molecule of water moving from stock to stock. One hundred million years ago, some of the water in Marie's Pond was, perhaps, a Tyrannosaurus rex for a month. Last week, some of it might have been a rosebush for a day.

The world looks a little different when seen from this perspective, doesn't it? Through the lens of stocks you can begin to see how everything in the world is either collected—whether for seconds or centuries—in ponds or on the move between them. Some of these movements from stock to stock are ancient biological cycles, like the migration of monarch butterflies between winter and summer homes. Some are part of the daily rhythms of modern societies, like people commuting back and forth between suburb and city twice a day.

Some movements between stocks cause health crises, as when stocks of coal in the ground are excavated and transported to become stocks of coal in a power plant. When the coal is burned some of it joins stocks of air pollution particles in the atmosphere. Prevailing winds carry some of these particles into children's lungs, where they become stocks that can exacerbate asthma and other diseases.

Once you see the world in terms of stocks, you might see yourself a little differently too. It's not an exaggeration to say that you are water and wheat and ocean minerals walking around, admiring sunsets, and reading books about systems and multisolving. "You are what you eat" is not a metaphor. You are also what you breathe. When it comes to chemicals that can pass through the skin, you are what you touch. Seen this way, pollution isn't something that happens around us, somewhere out there in the world, it is something that happens to us, literally.

Learning about stocks, you also understand why my mentor, the systems thinker Donella Meadows, cautioned us to be careful about assuming about the idea of some distant "away."[1] Materials might move off into a distant stock, out of sight, with little direct impact on you, but they don't go away. The stock of carbon in coal doesn't go away when it is burned; it just turns to a stock of invisible gas. The trash you threw away might disappear from the can you brought to the curb on pickup day, but it joins a stock at a landfill. Even the elements of our own bodies don't vanish when we die but instead move on and on through decomposition into the bodies of microbes and into stocks in the soil, water, and atmosphere.

Stocks Smooth Out Shocks

Stocks help keep systems steady. In fact, that is probably the reason Mr. Curtis dug Marie's Pond in the first place. Collecting snowmelt, rainwater, and the early springtime flow into a pond could have been a way to protect his dairy farm against the risk of a dry summer. In the same way, keeping your pantry full is a hedge against times when you're too busy to shop or the more serious risk that supply chain disruptions could leave grocery stores with empty shelves. A system with insufficient stocks, or empty ones, is an unsteady system, prone to shocks and patterns of boom and bust.

Wealth is a stock. In the dominant economic system, one of its primary purposes is to smooth out shocks and disruptions. Your stock of wealth influences whether engine trouble with your car throws your life into chaos or poses only a minor inconvenience. With enough savings you can afford to rent a car or special order a spare part. You can take a day off work to deal with the problem. Similarly, the size of a business's reserves influences its ability to weather a drop in sales. A country's reserves—of food, oil, and medicine—determine how it fares in natural disasters.

The stock of organic material in the soil helps it retain water, which can keep crops alive during a dry spell. A stock of spare sheets and towels comes in handy when unexpected company arrives. A stock of wisdom held in the lived experience of elders can be called on when situations a younger generation hasn't experienced reoccur.

Where stocks accumulate can be a matter of equity and justice. If wealth is concentrated in a small number of stocks, then those wealthy parts of the system will be more insulated from shocks.

The steadiness that stocks bring to systems is a double-edged sword. If things are good, stocks help them stay good. But if things are bad, stocks help them stay that way too. If the stock of artificial chemicals in the tissues of my body is high and threatening my health, I likely can't change that threat overnight. That's especially true if the chemicals have a slow rate of turnover, like a pond with only the smallest trickle of an outflow. If the stock of distrust between neighbors is high, it might take years of changed behavior for them to trust each other again, if they ever do.

The slow-changing nature of stocks provides resistance to change overall. Sometimes this resistance to change is welcome, as when stocks allow us to ride out more intense storms or prevent shortages of key supplies. But, facing cascading problems, we also need aspects of our systems to change rapidly. We need the stock of rooftop solar panels to expand quickly. We need the stock of gas-guzzling internal combustion engines to be replaced rapidly by cleaner transportation options. We need the stock of poorly insulated buildings to be retrofitted to standards of high energy efficiency. We urgently need stocks of people in poverty to shrink.

Imagine another pond, but one where water represents a fleet of vehicles. How can the fleet become more efficient? Only by changing the makeup of the pond, right? Only by the water flowing into the pond (new vehicles purchased) replacing the water leaving the pond (old vehicles going to the scrap heap). Since each vehicle in the fleet might last a decade, turning over the water in such a pond is a slow process. For even longer-lived stocks, like electric power plants or apartment buildings, turnover can be even slower.

Andrew Jones, cofounder of Climate Interactive, advises governments around the world with computer simulations of complex systems, including the global energy system. He often helps leaders understand the implications of this slow-to-change nature of stocks with a thought experiment: "If the world's energy supply was 100 power plants, then at our current rate we only shut down 3 every year. We replace those 3 and add 3 more since we are growing our energy demand globally. So, we add 6 new plants each year. At current rates, about 2 of those 6 new plants are zero carbon. So, we started with 100 power plants, almost all using fossil fuels. At the end of the first year, we have 106 power plants, 104 of them using fossil energy." Clearly, this is a slow-to-change system!

Turning over stocks of leaders is a slow process too. The makeup of a city council changes with elections of new candidates, perhaps new candidates with new thinking, but the council as a whole changes slowly, as incumbents retire or are defeated.

It's not only physical stocks that need to change at an unprecedented rate. Tumultuous times and converging crises require new thinking about who we are, how we relate to one another, and what constitutes a good life. We are being called to unlearn patterns of thinking about race, gender, and class with no time to lose and lives and well-being hanging in the balance. The number of people holding a particular idea is a stock, one that changes with learning, unlearning, and also with the death of people who hold certain ideas and the arrival of a new generation with ideas of its own. Not only is our physical infrastructure limited in its speed of change, but cultures, mindsets, and beliefs take time to change as well.

As the urgency for change builds, the slow-to-change nature of stocks is a challenge that those trying to multisolve will need to grapple with again and again.

Multisolving in a World of Stocks

Careful attention to stocks will help you when multisolving. Learning to notice stocks and clarify how you would like to see them change can help shape your multisolving strategy.

Step back for a big picture view. Imagine you are sitting high on a mountain, looking down across a watery landscape of ponds, lakes, streams, puddles, and rivers. There are gigantic, placid lakes, holding a significant portion of the landscape's water and changing slowly. There are also tiny puddles, filled only when it rains and at constant risk of drying out. Little streams connect some of the ponds together. Some ponds have more than one stream running into them while others have only one inflow. Some drain out via churning rivers, some by little trickling brooks.

Maybe parts of the landscape seem healthy, but others could use attention. Perhaps the puddle off to the east is the only water supply for a whole valley full of wildlife, and it's about to evaporate. Maybe the largest lake looks ready to spill over its banks, potentially flooding a village.

None of us are the masters of entire hydrologic systems like this imaginary one. But using this scene as metaphor, we can look out across any array of interconnected stocks and understand how they fit together. In fact, just seeking this perspective is the first step toward multisolving. Which stocks do you influence? Which ones do you understand in detail? And who are the people who know the most about the stocks you understand less well? With questions like these, you are starting to think about the connectivity of systems and about allies and common interests.

Once you have a sense of the general landscape of stocks, there are a few other ways to think about stocks and the opportunities for multisolving that they present.

Ask which stocks are too high and which are too low. Ideally, the stock of hungry children in a community would be empty. The stock of topsoil would be quite high. The stock of particulate matter in the air would be low, and the stock of people with safe access to daily physical activity would be high. We hope that the stock of wind turbines is growing and that the stock of gas-guzzlers is falling.

The potential for multisolving exists any time a stock is at or approaching an unhealthy level. That's especially true if, from that mountaintop view, you can see other too-high or too-low stocks that could be adjusted using the same intervention.

If the stock of children suffering from asthma is high, that's an opportunity for multisolving. It could be lowered if the stock of air pollution was reduced. Implementing policies that remove polluting vehicles from roads or replace dirty electricity generation with cleaner sources would change the stock of air pollution and in turn the stock of children with asthma. Those policies would change other stocks too, perhaps the stocks of clean energy jobs or greenhouse gases. When one policy leads to beneficial adjustments in multiple stocks, that's multisolving. It's as simple as that!

Consider the behavior of important stocks over time. We've seen how stocks give systems stability and inertia. While this helps systems resist shocks, sometimes it can hide looming problems.

A multisolving project might bring together a community, fishing businesses, and ecologists, all of whom have a common interest in a stock: a healthy population of fish. The stock of fish is refilled by the birth of new fish and drained by both natural deaths and harvesting. How's the fishery doing? Is it thriving? Stable? Depleting over time? How might the project partners know?

They could dip a net in the sea and count how many fish they pull up. Lots of fish? Must be a healthy stock, right? Well, maybe. But what if last year a larger number of fish were harvested than hatched? Would you still say the stock was healthy for the long haul? What if, year after year, the same trend continued? The stock would be draining, perhaps slowly, perhaps precipitously. Understanding the behavior of the stock over time would give a better sense of this key element of the system than a one-time measurement would.

The same logic applies when the filling of a stock is the problem. You could measure how many parts per million of pollution permeate the air on any given day, but what's happening over time? Is the pollutant at a constant low level? Are its levels falling, indicating some progress being made on pollution? Maybe levels are rising, indicating that air pollution might be heading toward more dangerous levels if no corrective effort is made.

For important stocks whose level you are trying to keep within some healthy range, it's not enough to know how full the stock is at any one time. You need to know what direction that level is heading so that corrective action can be taken before it reaches intolerable levels. Once the levels hit those dangerous thresholds, it may not be possible to return to healthier levels before lasting damage is done.

Scan for equity in important stocks. Sometimes differences in the relative levels of stocks within systems make them unhealthy or misaligned with the values of the people living within them.

How does the stock of playground equipment compare across zip codes in a metro region? How does the stock of wealth differ by race or gender (or both) across a nation? How does the stock of people with access to electricity compare between rural and urban areas?

Making inequitable systems healthier requires interventions that tackle the factors that lead to accumulation in some places and scarcity in others. Multisolving projects can be designed to address these imbalances. We'll explore this in much more depth in chapter 10.

Beware: long-lived stocks reflect the past (including its values, beliefs, inequities, and power structures). The long-lived nature of stocks means that they often embody old ideas or a societal power distribution from the past. The legacy of redlining in the United States is an example. Redlining refers to the coordination of white-owned banks, real estate businesses, and city officials to confine Black people to the poorest neighborhoods with the fewest resources, which influenced the building stock, the urban tree stock, the highways stock, and more. Historically redlined (and therefore majority Black) neighborhoods to this day tend to have fewer trees per acre than whiter neighborhoods, which impacts nighttime temperatures, energy bills, mental health, property values, and more.[2] Although societal attitudes in a city may have progressed from the days of overt redlining, without addressing the way those inequities are embedded in long-lived stocks like roads, trees, and parks, the impacts of those past beliefs will persist in the present. Even if a city were to invest in planting the same number of trees in Black neighborhoods per year as in whiter ones starting today, think how long it would take to reach parity given that it takes decades for newly planted trees to reach maturity.

Creating equity between historically inequitable stocks can be an extremely slow proposition, unless you find ways to actively increase the speed of stock turnover or somehow redistribute the material contained within stocks. That's challenging with a stock like trees, which take a long time to reach maturity, but there are plenty of ideas for transformative change—from reparations for slavery in the United States to land redistribution policies—that can hasten the turnover of stocks to address societal inequities.

Expect inertia. Because most stocks can't change instantaneously, strategies to change the level of a stock, whether to build a needed capacity or stop the accumulation of a dangerous substance, can take time.

In multisolving efforts, there is often an urgent need to change the level of a stock as quickly as possible. We want to stop the decline in biodiversity and boost soil fertility levels and clear toxins out of the air, all as quickly as possible. We want to build understanding and trust and cohorts of trained leaders. But none of those stocks can change overnight (and stocks like long-lived infrastructure and entrenched beliefs are the slowest parts of any system to change). Amid our tangle of crises and emergencies, this truth about stocks is sobering.

Not only must we urgently change the levels of critically important stocks, but we must face inevitable delays and lags, limiting how quickly stocks can change in response to our efforts. That makes it important for both you and your project's stakeholders to have realistic expectations about progress toward your goals. Expect inertia and keep at it. Stocks may not change quickly, but they do change, and continued effort is the key.

Worry less about the absolute level of the stock and focus instead on the direction the level is moving. Is trust improving? Is soil carbon increasing, even if only by a fraction of your ultimate goal? Are young leaders gaining skills, even though they still have more to learn? A depleted fishery won't be overrun with fish immediately after your program starts, but does the data show, maybe for the first time in decades, more births than deaths? Positive answers to these questions are all signs of success and reasons to keep going.

Still, sometimes the sluggish rate of change of stocks presents existential problems. Sometimes we just cannot afford to wait for slow stock adjustments to play out. There are also strategies to accelerate the rate at which stocks change, and we will look at some of these strategies in the next few sections.

Train, educate, and organize to accelerate the turnover of stocks. To help create an ecological society, stocks of coal miners and oil rig operators will need to fall, while stocks of solar technicians, home weatherizers, and ecosystem restorers will need to increase. That's part of what many climate and labor activists call a Just Transition, a rapid phaseout of fossil fuels that leaves no community behind.[3] Policy supports will be needed for a Just Transition, especially if it is to be fast and smooth. Workers will need training in new industries. They'll need assurance that they won't lose their health care when they transition to another industry. Helping speed the movement of workers between these stocks will be critical to facilitating a rapid-enough transition.

Some stocks that delay change aren't physical but instead are widely held beliefs, worldviews, or ideas. The options to change them, however, are similar. You can try to hasten the uptake of new thinking with an educational program or a communications campaign. You can also try to usher out unhelpful ideas by highlighting their flaws. Investing in the capacity of society to learn can help speed the change of stocks of ideas.

Sometimes speeding up the pace of change starts with transforming who is making key decisions. This might take the form of power building or political organizing to change who is elected to the stocks of decision-makers, like city councils.

Organizations sometimes seek to speed the turnover of stocks as well. The skills of a team can slowly evolve over time. But if you need faster change, say to embed new understandings about racism or gender into your strategy, you might need a more active training program that involves both unlearning old habits and building new skills.

Repurpose stocks to propel change. Conscious repurposing is when you take an existing physical stock and align it toward different systems goals. Repurposing a stock can be much faster than building a new one. It can be less expensive and more frugal with resources. It doesn't change stocks so much as direct them toward different needs.

During the COVID-19 pandemic, people in rural towns like mine worried about how children who depended on school lunch would be able to access those lunches once classes moved online. In a matter of days, school officials in my town came up with a plan to repurpose school buses. Following existing school bus routes, drivers would make midday runs, but instead of picking up children, they would drop off lunches. Existing stocks—school buses, drivers, established routes, the school cafeteria kitchens—were quickly repurposed to address an emergent need.

What else might we repurpose as we face more interconnected crises? Could colleges in precarious financial states meet local needs for housing or become sites of cooperative learning communities? Could lawns be turned into gardens?

Retire long-lived stocks early to speed up transformation. It is natural to want our investments to last as long as possible, even admirable to squeeze the most utility out of the embodied materials and energy. But if you urgently need to replace the water in a pond, then you may need to drain it via direct intervention, so that it can refill with fresh water. In economies, that's the idea behind early retirement.

"Cash for clunkers" programs are a good example of how early retirement of stocks works. These programs offer car owners an incentive—sometimes an outright cash payment, sometimes assistance with purchasing a new, more efficient vehicle—in exchange for taking their older, less efficient, more polluting vehicle off the road. Similarly, there's an ongoing push in many states right now to replace gas appliances like stoves and water heaters with highly efficient electric ones—early retirement on a micro scale that, spread across millions of households, could help squeeze fossil fuels out of the energy mix.[4] Other examples of early retirement include coal-fired power plants mothballed before the end of their expected operating lifetime, or gas pipelines shut down not long after their establishment.

Another way to speed change in long-lived stocks is to retrofit the stuff within them. For instance, you may not be able to change the building stock of a city quickly enough to meet clean energy goals just by slowly replacing old buildings at the end of their lifetimes with new more efficient buildings; accelerated retirement—tearing down perfectly sound buildings to put up more efficient ones in their place—could be wasteful. Retrofitting provides an alternative way to transform a stock by upgrading it.

Sometimes the best way to change a system is to let a stock flow free. Across the world, dams are being dismantled from rivers to allow water to flow freely and to restore habitat for fish and other species. Dam removal is literally the releasing of a stock so that what has accumulated may disperse. But it's also a good metaphor; think of all the unhelpful "dams" holding back stocks of accumulation in various places in our societies and economies.

The wealth of billionaires is a stock. So is the collective wealth of white people, which in the United States is a much larger stock than the wealth of Black people. In fact, in 2021 the average wealth of white households in the US was $250,000 while the average wealth of Black households was $27,000.[5]

We know that disparities in wealth and power create other disparities, in health for instance. To truly create a healthy society requires creating an equitable society; creating an equitable society might require opening up some stocks and redistributing their contents to a wider swath of the system. For example, progressive taxes, which ensure that those with the most wealth contribute the most to the common good, move some wealth from a stock of individual ownership to support collective well-being.

Writers and artists are often advised not to "hoard" their best ideas for another day but to work with them, create with them, and trust that the stock of ideas will be replenished by their own creativity. Creativity, artists tell us, replenishes when allowed to flow and stagnates when held too tightly.

And while stocks can help buffer systems to stay steady and healthy during shocks, sometimes resilience comes from the ability of stocks to flow freely. For example, in the aftermath of disasters, people outside the disaster zone have an urge to help, donating money, food and water, and sometimes traveling hundreds or thousands of miles to repair roofs and clean out flooded homes. They are opening up stocks, letting resources flow, and helping others navigate tumultuous times.

Sometimes the most important question to ask about stocks is "what can I share?" What portion of accumulation can be released? How can finding just the right point of accumulation and releasing some of whatever is held there solve more than one problem at the same time?

These questions apply not only to things like money or supplies but also to intangible stocks such as power. Do you or your organization have a reputation, a platform, or the ear of decision-makers? What would it look like to let some of that stock flow somewhere else? Could that enable the system as a whole to shift toward a healthier state?

? Questions for Reflection

- Take a walk in the woods, through your neighborhood, or around your office building. Notice as many stocks as you can and list them on a piece of paper. Which are long lived? Which turn over quickly? Do any of them connect to the others? Is the level of any of the stocks you see causing problems; if so, how might that level need to change? If you want, share the results of your walk with a friend or colleague.
- Think of an important stock in your world. Is its level steady or changing? If it's changing, is it draining or filling? How fast?
- Does the world look or feel different when you view it as a collection of connected ponds?
- How might environmental regulations change if they took more account of interconnected stocks?
- What consequences of stock depletion or overfilling do you see around you? How could those stocks be restored or drained?
- What stocks are you building right now, whether physical projects, websites, communities of relationship, or bodies of knowledge? What stocks do you want to be building?

Flows Everywhere

Spring flows around here.
Snow melts and runs in rivulets
    across frozen paths.
Snow melts and trickles and rushes.
March snowmelt moves,
it flows,
it soaks in.
Muddy boots and soggy mittens,
a big wet dog who drips.
And all you can do is laugh
    and fetch the mop.
Chase the flow, mop it up,
    squeeze it out into the sink.
Still it flows,
into the ground,
toward the brook,
to the river,
seeking the great flowing sea.
The rains flow down, drenching.
Sugar Brook sings.
The birds return, first the redwings,
    then robins and bluebirds,

flying, afloat,
a flow in a South wind.
Maple sap flows up from the roots,
out of the taps into our
    buckets and tanks.
In the sugar house the fire roars
    and steam flows out,
Boiling into the dark,
    star-studded sky,
and sweetness thickens
    in the boiling pan.
Jar it up, store it away on a
    dark shelf in the pantry,
until one day,
one crisp October day,
you break open the seal,
tip the bottle over your
    breakfast plate,
and sweetness
flows everywhere.

CHAPTER THREE

Flows

There are two inflows into Marie's Pond. The first inflow comes from the deep springs through which water bubbles up from underground. Rainfall is the second inflow. There are also two outflows. One is a little stream, Sugar Brook, that exits the pond; the other is the evaporation that occurs when water molecules on the pond's surface vaporize into the air.

Without naming it, we've been talking about flow, the second element of systems, quite a bit already in this book. Flows are so connected to stocks that it is almost impossible to discuss stocks without at least passing reference to flows. Flows are what move "stuff" from stock to stock. Inflows fill stocks, and outflows drain them. In your multisolving efforts it will be as important to think about flows as it is stocks.

Remember how stocks are things that can usually be measured or counted? Flows are processes that move. While stocks are measured in terms of units of stuff (gallons, people, trees, micrograms), flows are measured in units of stuff per unit of time (gallons per day, people per month, trees per year, micrograms per second). If Marie's Pond were to be measured in gallons, the inflows and outflows would be measured in gallons per day, arriving by rain and from spring water and escaping downstream and to the air.

It's important to understand, as you learn to discern the stocks and flows around you, that their units must match. In other words, the outflow of a stock of potatoes can never be dollars per month, but it can be pounds of potatoes sold to the grocery store per month. That flow might generate cash, but the flow itself isn't cash. This principle is more obvious with physical stocks and flows, like ponds and the streams; your pond of water would never have a stream of frogs or pond lilies flowing out of it. The principle is less obvious with other less physical systems, but remember, even a long chain of stocks and flows will feature consistent units across its entire length.

Cooking is a flow that takes a stock of food from your pantry and moves it into a stock called dinner. A second outflow, distinct from cooking, carries the carrot peelings and apple cores into the stock of compost. Eating is yet another flow, one that moves dinner into *you*, from which, eventually, flows of digestion, metabolism, and excretion move the material out of you. What might the units be for these flows? Maybe pounds or grams per hour or, perhaps, calories per day. The units you choose depend on what you're trying to understand by studying the stocks and flows in the first place.

Because stocks can be filled with intangibles, like trust, knowledge, fear, and confidence, flows of such intangibles must also exist. Knowledge stocks grow with learning and decrease with forgetting. Trust is built by having trustworthy experiences and declines whenever trust is breached.

Flows are important to multisolving because they are how stocks change. Say the goal of your multisolving project is to improve the number of fish in a coastal waterway as part of larger ecological restoration. If you can accomplish it, you'll see multiple benefits: ecological sustainability, economic vitality for fishing businesses, and community well-being for coastal communities tied culturally and economically to the fish. All the possible ways to influence the fish population involve changing flows. If your actions led to an increase in the rate of egg laying or boosted the survival rate of juvenile fish, you'd be influencing the flow of births per year. If you introduced new fish from a hatchery, that would add a new flow, one you might call introductions per year. Or you could decrease the rate at which the stock of fish declines by instituting harvest limits. If over-predation by natural predators was a problem, you might look for a way to counteract that and slow the rate of predation per year. Both actions constitute slowing an outflow.

Picture again the watery landscape from the last chapter, but this time focus less on the ponds and lakes (the stocks of water) and instead visualize the flows, the thin little streams, wide and slow rivers, and rushing whitewater rapids. These are as essential to the behavior of the landscape as the lakes and ponds are. If the little stream begins to run faster and faster, the ponds upstream and downstream will change too. Maybe the upstream pond will dry up and the one downstream will overflow. What could happen if a new flow is created by some geologic force that carves a channel between two previously disconnected lakes? The new flow might change the levels of the lakes, draining one maybe, or filling the other. It might even change how long water stays in each lake, speeding up or slowing down the turnover process. Your multisolving efforts can function accordingly, thoughtfully changing flows to fundamentally change how a system operates.

To Change the Levels of Stocks, Change Flows

In her talks and presentations, Donella Meadows used to share a cartoon of a poor fellow standing knee-deep in a flooded basement, with water continuing to pour onto the floor from a faucet. Mop and bucket in hand, he was frantically trying to soak up the flood. In other words, though the knob to control the inflow was in plain sight, all his attention was focused on the outflow. The cartoon always elicited a laugh of familiarity. We've all been in situations like the poor "mop guy."

When you need to reduce the level of stock, don't just focus on increasing the outflow; look around and see if you can slow down the inflow too. Facing rising homelessness in your city? Well, certainly focus on connecting people to services that could help them find housing (increasing the outflow from the stock of people in need of housing). But also ask why people are losing their housing in the first place; try to understand the inflow of becoming homeless.

When you need to increase the level of the stock, don't just focus on increasing the inflow. Ask if there might be ways to decrease the outflow. If your bank account is dwindling, you can try to work more hours, increasing the inflow. But you can also try to cut back on expenses, stemming the outflow. Trying to reforest a hillside? Keep planting trees, for sure. But don't forget to look at ways to improve the survival rate of the trees that have already been planted. In the face of a staffing shortage don't just hire more workers; ask why you have such a high quitting rate in the first place.

Sometimes the goal of multisolving will be to increase a stock to improve a system's health. We might want to see more three-year-olds in preschool or more trees on the mountainside or more women in positions of political influence. At other times, system health might depend on stocks falling, such as lead in drinking water, greenhouse gases in the atmosphere, or dollars lost to white collar crime. And sometimes we want stocks to hold steady because we have just the right number of cows in the pasture, children in the program, or money in the bank. An understanding of flows tells you exactly what needs to happen to achieve each of these goals.

To **increase** the level of a stock, the sum of the inflows must be larger than the sum of the outflows. For my little pond's water level to rise, the total amount of rain entering the pond from rainfall and spring flow must be greater than the total amount of water leaving the pond via the stream and evaporation.

To **decrease** the level of a stock, the sum of the inflows must be less than the sum of the outflows. For the level of CO_2 in the atmosphere to begin to fall, the absorption rate of CO_2 into oceans and plants must be higher than the release of CO_2 from burning fossil fuels.

To hold the level of a stock **steady**, the sum of the inflows must equal the sum of the outflows. If we want our neighborhood association's rainy-day fund to stay steady, we need to make sure that annual expenditures, the outflow, are balanced by new funding and grants.

Whenever you multisolve, you will encounter stocks whose levels need to be changed and others whose levels would, ideally, hold steady. These three simple "laws of flows" will give you a quick way to assess your options for meeting your goals.

We've seen how stocks hold accumulations of stuff and how flows fill and drain those stocks. Stocks bring constancy to systems, and flows bring change. But in real life, systems are more complicated than that. Flow rates themselves can fluctuate.

A steady flow rate, all else being equal, means a steady change in the stocks it fills or drains. Vary the rate of flow, and the stock's behavior becomes more complex. When you change a flow rate you alter the rate at which change happens. This point is obvious when you consider flows of water. Sugar Brook, which drains Marie's Pond, flows swiftly in April when the snow melts. In September, after a long hot summer, it is a thin, slow trickle. The rate of its flow, the amount of water passing any given point in the stream per minute, is different in the spring and fall. If you're a farmer who uses Sugar Brook to irrigate your crops, you might wish you could make it flow faster in the late summer. If you're the mother of a toddler who likes to roam down to it, you might wish you could slow the water down in April. In both scenarios, if you couldn't change the rate of water flowing, you might choose to adapt your own behavior accordingly.

Flow rates are sometimes called rates of change. A freelancer's monthly income can fluctuate depending on how much work she has. The rate of change in her bank account is measured in dollars deposited per month. With fluctuating payments the rate of change itself is changing, month to month.

In tumultuous times, odds are that important flows—important rates of change—are themselves changing. The flow of carbon dioxide pollution from human activity, measured in gigatons of CO_2 per year, isn't constant. It has been rising since the Industrial Revolution. The flow of people migrating in response to climate impacts and the number of acres burned by wildfire per year have been increasing too. In response to these changing rates of change, the number of people joining the climate movement each year has been changing too. Changes in rates of change can lead to changes in other rates of change elsewhere in a complex system!

Sometimes we see steady increases in rates of change, like the gradual increase in CO_2 emissions per year. Sometimes we see sudden perturbations, like those caused by economic shocks or natural disasters. Whether gradual or sudden, changes in the rates of change can give rise to different, and sometimes unexpected, systems behavior.

One summer my husband, Phil, and I decided to hike a trail we'd never explored before. According to the guidebook, it was supposed to be an easy three-mile trip around a pond. If it had been a more strenuous hike, we would have left our dog, Thea, at home. But a three-mile flat stroll seemed like something she could handle, even in her old age.

The first two miles of the hike were exactly as advertised. The trail was wide enough that we could walk side by side, and the grade was gently rolling, never too steep. Then, however, the path narrowed and got wetter and wetter. Soon we were crouching down single file, pushing our way through branches. After a few more minutes, the path wasn't simply muddy, it was a little trail-shaped waterway! We pulled out our map and confirmed that we hadn't made a wrong turn. We were in the right place. But where we expected an open trail crossing a narrow stream, we were in the middle of a swamp.

We forged ahead. After a short distance the mystery explained itself in the form of freshly gnawed branches. Beavers! In fact, right ahead of us was a well-built beaver dam. A few minutes, two pairs of muddy boots, and one very wet dog later, we made it out. After balancing across the top of one beaver dam and thrashing our way through some deep pools of standing water, we clambered up onto dry ground and finished our hike.

Beavers change the rate of flow of streams by building dams, which impede the flow of water, causing it to back up into slow-moving pools. In so doing, beavers affect entire landscapes. In recent years, scientists have come to understand how much beavers change the hydrology of ecosystems.[1] Their dams create meadows and wetlands. They also provide downstream flood control by slowing the flow of water after heavy rains, allowing it to percolate gradually down into the water table.

What beavers have always done is now becoming a principle of urban design in the era of climate impacts. "Soften the city" is becoming a new mantra. Urban planners are replacing concrete and asphalt with permeable pavement, rain gardens, green roofs, and urban wetlands. Each of these actions changes the rate of surface water's flow across the urban environment. Without slowing, the water moves fast, becoming what hydrologists call "flashy." Fast-moving water across hard surfaces makes neighborhoods prone to flash flooding. Slowing the water, allowing it to soak and spread, protects people and property.

Both the beavers and the urban planners show us that changing the rate of flows changes systems. The beavers changed the rate of one flow—the speed of water moving through the system—and look at what changed in response: the shape of the landscape, the accessibility of the trail, the mix of habitats, the risk of flooding far downstream. A single change to one flow transformed the whole system.

What if the annual allocation to a city's emergency management department increased by 50%? What if the annual flow of CO_2 into the atmosphere from the burning of fossil fuels fell 90% compared to today? What if the number of cars per day driving in the city center was cut in half? Changes in flow rates can drastically alter the behavior of systems, making flows enticing targets for multisolving.

The Flow Perspective

Stocks are what you tend to see when you look at a system: physical and concrete, countable. Buildings, cars, planes. Trees, wolves, e-bikes, solar installers. Stocks draw your attention, but flows are where the action is. Stocks give momentum and stability and inertia; flows give dynamism. Flows are where the life and movement and change are to be found.

Our culture's focus on stocks and neglect of flows is to our collective detriment. It keeps us from seeing reality clearly, starting with the reality of our own bodies and our own minds. Right now, sitting here at my keyboard in my writing cabin, I look down at my fingers as they type. Long, summer brown, fingernails dirty from the vegetable garden, a little swollen at the knuckles. My fingers look like things, and of course they are. They appear permanent. There's the scar where a pocketknife, illicitly procured, snapped shut on my right index finger more than forty years ago. Fingers are things of some significant duration.

But the flows through them—oh my goodness! The carbohydrates from my granola for breakfast are breaking down as I type, burned up to power the muscles that lift my fingers and move them from key to key. That's a flow of energy and of matter.

Despite the scar from that pocketknife mishap when I was eight, these are not the same hands from all those years ago. That's obvious in the newer scars, the blueberry picking stains, the knuckles that are starting to swell just a bit with age and maybe some early arthritis. But in fact, these hands aren't even the same hands as yesterday. The skin that covers them is constantly sloughing off and being replaced. The molecules of collagen and keratin that were "me" last week are not me today, and when I meet an old friend I haven't seen for years, her face is at once instantly recognizable and, literally, new. We stay the same by constantly changing.

These hands haven't become something entirely different; they're still hands. But like flames or whirlpools or spiral galaxies, they both take in and release matter and energy in a dance that will not stop until the day I die. Actually, the dance won't stop even then; it will merely change form, the energy and matter that was once "me" animating soil bacteria or insects.

All of that is "just" my body. What about my mind? My stock of knowledge? I added to it yesterday when I learned the location of a new swimming hole and mastered a tricky new knitting stitch. My stock of experience is growing too. I had never lived through a global pandemic until 2020, nor had I gone so many months without hugging my daughter. Not only am I not the same person I was physically a few months ago, my stocks of memories, knowledge and experience, my level of skill, my confidence in certain ideas and beliefs, these are all changing day by day, minute to minute, thanks to flows of learning and flows of forgetting and unlearning. And I'm just one eight-billionth of human beings here on Earth. Think of all the combinations of us. Families. Neighborhoods. Cultural diasporas, not all in the same place, but still connected. Nations. Businesses. We exist within stocks that hold their shape even while constantly changing.

For more than twenty years I've lived in in co-housing community in a small Vermont town. In those years the size of the community has stayed roughly the same—twenty-three households, around thirty-five adults—even though almost half of the founding families are no longer my neighbors. Many have moved away; a few have died. The whole of the community—our risk tolerance, our project priorities, our level of conflict and our ability to solve them—all of these attributes shift as new members bring new perspectives and experiences, and departing members leave some holes. No one writes funny songs quite like my neighbor Phil did, and no one has the patience to light hundreds of luminarias for the harvest celebration the way my neighbor Jay did. But now Helen and Sandy inspired a ukulele group, and Howard has endless enthusiasm to maintain walking trails through the forest. With every coming and going, the whole changes.

Multisolving is possible because of this constant dance of change. When you multisolve you are working with all this flow and movement, nudging it, steering it, creating tiny new rivulets or giant rivers, or plugging holes to stop something vital from leaking away.

If our decisions don't grapple with the reality of flows, they may eventually create problems. Agricultural policy that ignores the flows that connect pesticides sprayed on apples to a baby's developing organs could contribute to health problems. Highway design that doesn't consider flows of animals across roads in their search for food and mates can undermine efforts to protect biodiversity. Unemployment policy that doesn't offer a flow of compensation that matches the flow of lost weekly income can push people into hunger, utility shutoffs, and evictions.

You can neglect imbalances of flows for a while, but eventually, if inflow exceeds outflow or vice versa, stocks will drain to zero or overflow. If that overflowing or emptying is consequential enough, you'll have a crisis. The overflowing stock of CO_2 in the atmosphere is a climate crisis. The draining of aquifers of fossil water is an agriculture crisis.

If, year after year, the inflow of investment in bridge repair in your state is less than the outflow of bridge wear and tear, then you will experience decline in the structural integrity of bridges. If the number of nurses graduating from nursing school each year falls short of the rate of nurses retiring, the nation's health care sector will eventually find itself in a staffing crisis. If plastics are being thrown away faster than they can be properly disposed of or decompose (a very slow rate indeed), plastic waste will accumulate. But that plastics problem is just a symptom of a deeper problem, an inability to see and work with the flows.

For all these reasons, many multisolving efforts find themselves focused on changing the balance of inflows and outflows.

Multisolving in a World of Flows

Some processes create multiple flows. This common feature of systems opens up multisolving possibilities. For example, the combustion of fossil fuels creates a flow of both CO_2 and small particles that contribute to air pollution. If burning fossil fuels could be curtailed, both the CO_2 flow and the air pollution flow (and all the problems that stem from each) would diminish too.

Investments that make a city more walkable will influence two flows right away. The flow of walkers per day will go up while the flow of drivers per day will go down. Other consequences will spiral out from those two changed flows. Part of the stock of people who are sedentary will flow into a stock of people who get healthy amounts of physical activity. Health outcomes will result from that, which will change flows of spending in the health system. Meanwhile, flows of emissions and pollution from transportation will decrease, which will also impact health, and health spending, and climate change. And so on. There are a few more flows that are likely to change because of these changes. See if you can think of a few of them.

If you could slow the flow of plastic, you could multisolve for ocean health and the health of people living near the polluting facilities where plastics are made. If you were strategic about how to slow the flow from farm harvest to landfills and increase the flow to community kitchens, you could multisolve for food security, climate protection, and economic opportunity. If you could increase the flow of homes per day switching from gas boilers to heat pumps powered by clean electricity, you'd be fighting climate pollution, improving indoor air quality, and supporting local businesses.

Many possibilities for multisolving emerge from influencing flows, so let's look deeper into how to work with flows.

Track the relative inflows and outflows of important stocks. If you want to understand how a stock is going to behave over time, look at the relative size of inflows and outflows.

If the concentration of a pollutant is increasing, the rate at which it is being dumped into the environment will be higher than the rate at which natural systems can render it harmless. A target of your multisolving might be to bring out-of-balance flows back into balance.

If the rate of new housing being built in your city (the inflow) exceeds the rate at which old housing is being demolished (the outflow), the stock will be growing. If your multisolving goals include establishing more housing, you can support the inflow, try to slow the outflow, or both.

If your goal is to expand the membership of a movement, what needs to be true about recruitment versus people leaving the movement? You would need more people to join than leave, right? And you could work on the inflow (more outreach) or the outflow (less quitting) or both.

Multisolving can involve creating new flows or restoring existing ones. Sometimes to restore system health, a flow that's defunct or in decline—such as a river that has been diverted for agriculture—needs to be restored. Or a healthier system might require new or larger flows of resources. Think about wealth accumulated over time by an institution from the exploitation of people or natural resources. How could that wealth flow again? From reparations for slavery to land redistribution to investing in ecological restoration, we can encourage stocks to flow again.

If a new or increased flow can serve multiple goals, it's a multisolving strategy. In Portland, Oregon, for example, campaigners worked together to build support for a 1% clean energy surcharge on retail sales by large corporations.[2] That flow (an estimated $60 million a year) is being used to support energy efficiency retrofits and training for clean energy jobs. The new flow is helping save money for residents, reduce greenhouse gas emissions, and increase access to good jobs.

Reducing or eliminating flows can be multisolving too. Adding a new flow isn't the only way to make a system healthier. Sometimes what's called for is reducing or even eliminating a flow. When an ecosystem is being overharvested relative to its regeneration rate, the harvesting flow needs to be slowed down. When topsoil is eroding faster than new soil is being created, an erosion reduction program is needed.

Whenever something essential for system health is leaking away or failing to accumulate as you expected, look for an outflow that might be running too high. If the consequences of that overactive outflow are impacting several parts of a system, the conditions could be ripe for multisolving. Mitigating soil erosion, for example, protects future food security, land fertility, farmers' budgets, and downstream water quality, all at the same time.

Be creative. Many organizations and governments have processes to control flow rates. If you're involved in these processes, challenge yourself to identify the flows involved. Can you name the flows under discussion? What stocks do they influence? What other flows are upstream or downstream? How have the flows behaved in the past, and how might they change under future stresses and risks?

Policy changes can influence flows. Can we pressure the federal government to protect more acres of land per year? Can we challenge the designers of the new bike path so that it can accommodate more trips per day? Can we influence the board to commit to doubling the number of patients the clinic sees each year?

Investment decisions can change flows. When the school board increases the budget line item for library books, more books will be purchased per year. When an ecosystem restoration project invests in new tools or more workers, it increases the number of acres that can be restored per year.

Investments that support people also change flows, including the investments we make in ourselves. With practice, the rate at which I can stack wood, fold laundry, knit sweaters, or write an op-ed all increase. A skilled manager can mediate conflict and support team cohesion, which in turn increases the team's flow of work accomplished per month.

Sometimes a flow's rate is limited by ancillary factors. In a garden, for instance, the growth rate of plants may be limited by one missing cofactor or nutrient. The gardener may be able to boost the growth rate of plants and the productivity of her garden by supplying that nutrient.

If providing support changes flows, taking support away does too. Workers who lack tools and resources are likely to be slower at getting their jobs done. Ecosystems that are starved of a key resource may be less efficient at capturing solar energy.

Collaborate. Setting the "dial" on key flows can be a point of contention or even conflict. There may be multiple forces within a system, each with their own sense of the desired rate for a particular flow, pushing in opposite directions. It can take power to change flow rates, like the political power to pass a policy.

If a system seems stuck, it may be obvious which flow needs to change. But odds are that if changing the flow were easy, it would have already happened. To accomplish your multisolving goals you may need to prevail in a power struggle to determine the rate of a particular flow. That's where multisolving proves to be an advantage. If the flow in question ripples out to create multiple problems, then each constituency impacted by that flow is a potential ally in changing it.

If the flow of stormwater is flooding a neighborhood and straining the city's sewage treatment facility, residents and city officials might collaborate on installing rain gardens and green roofs to soak up and slow the flow of water. Such a change in flow could also solve other problems, like unemployment (by providing good jobs for youth) or heat islands (by keeping the neighborhood shady and cool).

Seek the big picture when flows change. Sometimes changes in rates of flow are welcome developments. Nevertheless, adjusting to changes in flow rates that have existed for a long time, to which we are accustomed, can be uncomfortable. It can take some getting used to changes even when they're desired. Stop and reflect if flow rates are changing in a system of which you're a part. Sometimes noticing and acknowledging discomfort with change is a way to lessen it.

Your city's plan for reducing annual greenhouse gases from transportation (a flow) might mean you can no longer afford the parking fees downtown and have to take the train after years of commuting by car (you'd have left one flow [of cars] and joined another [of train commuters]). That could feel disorienting. You might miss your old familiar routine, the quiet thirty minutes during which you planned your day. System change even when we know it's essential or for a good cause, even if we have fought for it, can be uncomfortable.

For societies to become more equitable, many flows will need to change. In the United States, the flow of new admissions to law school per year might once have been made up largely of white men. In a more equitable society, that flow will better reflect the demographics of the country. For many that change will feel like an opening of opportunity. But if you are a white man, a group that has had a disproportionate share of the flow, the change could feel disconcerting.

In moments of discomfort, take a step back. Try to see a larger view. Look at all the flows and at their historical patterns. In systems undergirded by patriarchy or white supremacy, flows will be out of balance. They might be actively held out of balance by power, privilege, or even violence.

Remember that view of stocks and flows from the top of the mountain? In inequitable systems some flows are far bigger than is equitable, and others will be trickles. It helps if you can find the perspective to see beyond your individual experience to see the entire landscape. Your individual experience might be *Hey! This flow I'm in is shrinking*, even though from the mountaintop perspective, you can see that the flow is still strong and vigorous. It's just more in balance now with other flows moving in the same direction.

So far in this book, we've looked at the rates of flows in isolation from their larger systems, but of course few flows exist in isolation. Systems really become interesting when we start to consider their interconnections.

What happens when a flow fills a stock that also influences the rate of that same flow? In fact, this is a common phenomenon in systems, including our bodies. The flow of eating fills my stomach, which makes me feel full and diminishes my rate of eating.

In geopolitical systems, a country might gain power via a flow of conquest or by the economic domination of its neighbors. If it then deploys that stock of power to gain even more power, the change in the flow of gaining power feeds back to increase itself. This, of course, is part of how empires are created.

When a change in one part of a system feeds back to change that same part, we call that a feedback loop. In the next two chapters, we'll look at two types of feedback loops, both of which present possibilities for—and obstacles to—multisolving.

? Questions for Reflection

- Where could slowing a flow improve the health of systems in which you participate? Where might increasing a flow help meet important goals?
- Explain the three laws of flows to a friend or family member.
- What are some of the most (geographically) distant flows that impact your world? What are some of the closest/most immediate?
- Pick an important flow in your life (or several). Whose hands (literally or metaphorically) are grasping for control of its rate? What power would you need to move that flow rate to the level you'd really like to see?
- For each of the following changes in flows, identify some of the benefits that might be created and name a constituency or interest who would care about that benefit. Each of these interests represents a potential multisolving partner toward making that change happen.
    - Reducing the flow of food waste from restaurant kitchens to landfills
    - Increasing the flow of successful hatching for native pollinators in a farming region
    - Decreasing the flow of toxic releases from a petrochemical factory
- You can try this question for any flows of importance in your own world!

How to Make a Giant Snowball

One. Gather some snow.
Wet snow. Sticky snow.
If it is dry, if it is snow with
*   no affinity for itself,*
*   well… I don't know*
*   what to tell you.*
Stickiness matters is
*   what I'm saying.*

Two. Pack your snow
*   round and tight.*
A perfect little round ball.
Do not worry if it is small.
The size of a grape, a
*   kumquat. It's fine.*

Three. Carry your
*   snow-kumquat up a tall hill.*
The steeper the better.

Four. Place your small, sticky
*   snowball on the ground.*

Five. Give it a push. Hard or
*   gentle, it's fine. Watch.*
Scream or shout, or steal your
*   brother's hat if you like.*

Six. Run downhill.
*   Or roll or go swooshing*
*   on your shiny red sled.*

Seven. Collect your giant snowball.

Eight. Do it again, if you like.

CHAPTER FOUR

Reinforcing Feedback

Flows can be turned up or down based on the state of other variables within a system. Stream flow might change based on a stock of beavers in the watershed. A flow of dollars into my bank account might change based on a stock called my reputation as a consultant. The flow of gallons of milk off the grocery store shelves might increase as a stock called hype about the upcoming blizzard increases.

Not only do changes in stocks change rates of flows in general; sometimes a change in the level of a stock can influence its own inflow or outflow. Suddenly, cause and effect are no longer unidirectional, and you're witnessing a loop of circular causation: a feedback loop!

Linear causation is when a change in A causes a change in B. Circular causation is when a change in A causes a change in B that causes another change in A. Feedback loops are a basic feature of complex systems.

For instance, when the balance in your savings account increases, the flow into your savings account called interest accrual increases as well. The size of your savings account feeds back to change the inflow into your savings account.

Your level of fitness is also a stock. When it increases, due to an increase in your daily physical activity, you might feel more inclined to take on more jogging or more challenging hikes. Your level of fitness increases the inflow of improving fitness that comes with exercise. Up to a point, the more fit you get, the easier it is to get fitter.

A salesman with a goal to make a certain number of sales per month is operating according to a feedback loop as well. If his sales per week have him off track for meeting his goal, he might increase his number of sales calls the next week to catch up. The low level of a stock of sales increases the pressure to increase the inflow into that stock.

Feedback loops interact with one another too. One stock might be part of several feedback loops. Those loops may operate with different strengths and at different speeds. Variables increase or decrease under the net influence of all these spinning loops. Feedback loops make system behavior complex and unpredictable and can cause crises to converge and amplify one another.

But this complexity also creates the possibility of multisolving. When one variable in a system is part of several feedback loops, change in one part of a system can ripple through to others. Seeing feedback loops helps you see possibilities for multisolving. You can also use your awareness of feedback loops to design your change strategies.

In this chapter and the one that follows, we'll investigate the dynamics of feedback so that you can anticipate it, live with it, and take advantage of its power.

———— ∞ ————

There are two types of feedback loops.

The first two examples in the previous section—the bank account whose inflow increases as the stock of savings increases and the physical fitness stock that becomes easier to increase once it begins to grow—are examples of what is known as *reinforcing* feedback. In both cases, a change in one part of the system feeds back to amplify the initial change. That's true of all reinforcing feedback. Change creates more change in the same direction.

The final example—where a lower than desired level of a stock increases efforts to fill the stock—is an example of *balancing* feedback. While reinforcing feedback amplifies change, balancing is goal seeking and tends to dampen change.

We will focus on reinforcing feedback in this chapter and on balancing feedback in the next. So, let's delve into how reinforcing feedback works. What behaviors does it create in systems? How can we use our understanding of reinforcing feedback to better implement multisolving?

Recognizing Reinforcing Feedback Loops

Reinforcing feedback is change that feeds on itself. A small increase feeds back to cause a larger increase. A small decrease feeds back to create a larger decrease. When someone says that "small changes can snowball," says that something "spread like wildfire," or refers to a "downward spiral," they're talking about reinforcing feedback.

Pop culture fads, including fashion, are a great example of a reinforcing feedback loop. Fads spread, seemingly overnight, driven by word of mouth and the influence of "trendsetters." As more people adopt the fad, more people desire to participate, which leads to even more people adopting it, and so on.

If you've ever watched a social media post go viral, you've seen this happen in real time. The more a post is shared, the more likely it is to be shared even more widely. That's reinforcing feedback. For that matter, consider the spread of an actual virus. As the number of infected people increases, the number of exposed people increases, which results in the number of newly infected people increasing as well. An increase in the number of infected people leads to an even larger increase in the number of infected people.

Precipitous declines are also driven by reinforcing feedback, as when a declining stock of cash on hand leads to a decline in rates of investment and repair, which could lead to less productivity and even less repair.

Over time, reinforcing feedback creates a characteristic pattern, one that you can recognize on a graph but also in your gut. When things are changing rapidly, when a situation feels out of control—like an explosion, wildfire, snowballs rolling downhill, cancer spreading, collapse—look around for one or more reinforcing feedback loops driving the process.

On a graph, reinforcing feedback that is growing (i.e., exponential growth) looks like a hockey stick—a long, slow incline that suddenly, abruptly shoots upward. Imagine a graph that charts the number of shares of a viral social media post per hour. At first, there would be relatively few shares. Then, as sharing leads to more people seeing the content, more people decide to share it, and the number of shares would trend upward. In fact, that number would increase at a faster and faster rate each time you measured. The upward trend of the graph would get steeper and steeper, until all the way at the right of your graph, you might see an almost vertical line, shooting straight up.

The trends generated by reinforcing feedback loops tend to fool our intuition. In the early stages, change can be so gradual that we fail to notice as it compounds on itself. Later, we are shocked by what seems like a sudden and drastic change. Where did that come from, that explosion of extreme weather or disappearing species or resentment from a coworker? Our systems can manage the situation during the long slow buildup but can get overwhelmed by the seemingly sudden explosive change.

Many of the crises that cry out for multisolving exhibit the pattern caused by exponential growth. They have been growing for a while but suddenly become unmanageable. The number of coastal sites where low oxygen levels have been reported around the world has grown in this sort of pattern, for example.[1] The number of people displaced each year by persecution, conflict, violence, or human rights violations shows this pattern.[2] So does production of plastics and global water use.[3] All these trends have been steadily building, but then, suddenly (or so it feels) we are confronted with the full magnitude of exponentially growing change.

But reinforcing feedback loops can fool our intuition in another way: sometimes we underestimate the possibilities for solutions. Social movements also grow by reinforcing feedback. The more people who participate, the more visible the movement becomes and in turn the more people are prompted to join. That's a reinforcing feedback loop.

New technologies spread via reinforcing feedback. The more people who install a new technology, the greater the awareness of it becomes, and the more likely new customers are to give it a try. The more units are manufactured, the more the price falls, leading to more demand and even more manufacturing.

If we expect steady linear growth from movements or technologies, it may seem that our efforts will never be large enough to make a difference. But just as problems can seem to explode from nowhere, solutions can too.

Stopping or Slowing Reinforcing Feedback

When it is unregulated, reinforcing feedback tends to create dysfunction in systems. So, bringing reinforcing feedback loops back under control can be an effective multisolving intervention.

In healthy systems, reinforcing feedback is regulated by other feedback loops that help hold it in check. We recognize the breakdown of those regulatory mechanisms as illness or disease, as something out of control or out of balance.

In your body, cells grow by division, which is controlled by a set of redundant biochemical pathways. One cell becomes two, two become four, four become eight, and so on. Cell division is essential to life. But controlling that growth is essential too. When cellular growth control breaks down, it results in illnesses, including cancer.[4]

Field mice populations grow via a similar reinforcing feedback loop of reproduction. Predators, including coyotes and owls, keep them in check. If habitat loss, disease, or pesticides reduce the predator population, the mice population can grow out of control. When reinforcing feedback slips free of regulation like this, it can ripple out into systems in disruptive and dangerous ways.

The loss of regulation of reinforcing feedback creates dysfunction in nonbiological systems as well. Power can grow via reinforcing feedback.

Those who have power can use it to write economic and political rules to favor themselves, thus gaining even more power. Democratic systems of governance recognize that this is a dangerous feedback loop. Term limits, elections, a free press, and campaign finance rules are all meant to keep any one individual or group from gaining unchecked power. When these weaken, democracies falter.

If important systems in your world are spinning out of control, look for reinforcing feedback loops. Look for places where change is feeding on itself in ever-escalating cycles. Ask yourself what regulates those feedback loops, or used to. Finding ways to strengthen those weakened spots is a potential target for multisolving.

***

Some opportunities to multisolve may involve interrupting a reinforcing feedback loop.

Consider how highway design impacts transportation policy and outcomes. All else being equal, creating more traffic lanes makes driving more attractive, which leads to more cars, requiring more lanes of roadway to relieve congestion, making it more attractive to drive . . . ad infinitum. One change (building more lanes) feeds on itself to create more change in the same direction (building even more lanes). This reinforcing feedback loop exacerbates congestion, air pollution, noise pollution, safety hazards, and greenhouse gas accumulation. With so many problems created by this one reinforcing feedback loop, any success in slowing, blocking, or reversing it would multisolve. Interventions might shift public investments away from roads and toward other modes of transportation or institute congestion pricing to make driving into the city center less attractive.

Inequities can persist or grow because of reinforcing feedback loops. When wealth and power end up in the hands of a small group who use that wealth and power to influence the making of rules, laws, or incentives, that's a reinforcing feedback loop. Having power increases the odds of having even more power in the future. Since inequities and disparities give rise to all sorts of other problems—including health and economic impacts—finding ways to slow or reverse this reinforcing feedback loop represents a huge multisolving opportunity.[5]

When investors invest in capital stocks designed to extract resources from nature (fishing boats, sawmills, oil wells, etc.), the resulting extraction produces profits. Some of those profits are then reinvested in more extractive capital. This reinforcing feedback loop can drive economies beyond the limits of local ecosystems. Interventions that slow or stop this feedback loop are multisolving. Relieving the pressure on planetary life support systems is a fundamental intervention from which many ripples of solutions emanate. Efforts to reduce consumption, establish circular economies, and enact harvest and extraction limits are all examples of regulating this growth of capital feedback loop.

The following are a few simple principles about interrupting reinforcing feedback.

Sometimes you must adapt to the symptoms created by reinforcing feedback but intervene in the loop itself too, if you can. In the face of problems driven by reinforcing feedback, you may need to put energy into adapting. To protect people and places you may well need to adjust to shifting conditions, whether those are new coastlines or higher concentrations of a pollutant. Adapting to impacts from reinforcing feedback may save lives, keep an organization functioning, or help maintain people's quality of life.

Adapting to changes driven by reinforcing feedback can be done in the spirit of multisolving. New infrastructure to resist climate impacts, such as increasingly dangerous heat, can be built in ways that provide workers with good jobs and communities with new amenities like parks and green roofs and their attendant health benefits. But don't be fooled. Unless something disrupts the underlying engine of reinforcing feedback driving those rising temperatures, the problem will not only get worse but worsen at a faster and faster rate.

The capacity to adapt and adjust to most crises does not grow exponentially. While the number of people infected with a virus can climb and climb, the number of hospital beds most likely cannot. While the population of an endangered species might suffer exponential decline, the budget for ecosystem restoration will probably not grow accordingly. For most problems driven by reinforcing feedback loops, the problem can outpace a system's ability to cope. Adapting and adjusting can buy time and prevent losses. But to succeed you must ultimately find a way to slow or stop the reinforcing feedback.

Reinforcing feedback loops are large, overpowering, and sometimes frightening by their very nature. It is tempting to try to find ways to live with them, to compensate. But this provides only a temporary illusion of safety. So adapt and adjust to take care of people and nature as you can. But make sure that somewhere in your strategy is a plan for intervening in the reinforcing feedback loop itself.

Slow down the rate at which reinforcing feedback creates multiple problems. The strength with which change amplifies with each cycle around a feedback loop is known as its "gain." The higher the gain, the faster change feeds on itself. If a reinforcing feedback loop is creating a crisis, multisolving may require turning down the gain.

Imagine you're a firefighter. To successfully put out a fire you need to starve it of either fuel or oxygen. Now, for whatever problem you are facing in real life, what fuels its growth? What is the analog for oxygen? And how can you interfere to put out the metaphorical fire?

If misinformation is spreading, can you debunk it? If profits from environmental extraction are being funneled directly back into more extraction, can you somehow turn down that reinvestment rate? Taxing extraction and using funds to restore the resource would do the trick.

If a reinforcing feedback loop is causing multiple problems—like depleting a resource, creating pollution, and harming communities all at once— slowing the loop down will relieve the pressure on multiple parts of the system simultaneously. That's the essence of multisolving.

Act as early as you can when reinforcing feedback is at the root of your need to multisolve. Early on in a reinforcing feedback process, small changes result in only a little impact, but after a while each trip around the loop becomes more extreme.

If you know that eventually you're going to need to intervene in a problem that is growing exponentially, don't wait. If democracy is being overrun by a power grab, act early while that gathering power is still weak. If the growth of an industry is overrunning communities or ecologies, limit that growth early if you can. The easiest time to stop a runaway train is just after it starts moving. The next easiest time? A few seconds later.

When it comes to reinforcing feedback loops, "wait and see" is almost always a bad idea, whether responding to a house fire or climate change. But although early intervention is for the best, it is not always easy to be the intervener. Be warned: observers who do not understand the potential for a problem to escalate may view your swift action as disproportionate to the situation. "It's just a little problem, what are you so agitated about anyway?" Whether that's your neighbor questioning your passion for climate action or a political leader shrugging off the danger at the early stages of a pandemic, those who aren't focused on the power of reinforcing feedback may paint those who are as "alarmist."

Try to help those advocating for a slower approach understand that your response is calibrated to where the system might take you. You are responding to the momentum and the dynamics, not to the problem's current condition. To recruit others to your cause of early action, it may help to describe the loop at play using analogies to more familiar reinforcing feedback problems like wildfire or cancer.

Anticipate the momentum of reinforcing feedback. Most people tend to focus on the symptoms in a system. How many fires are burning in California? How many children are going to bed hungry? What fraction of the staff are feeling angry or disenfranchised?

Understanding reinforcing feedback allows you to look beyond the current conditions of the system to anticipate its momentum. If you identify a reinforcing feedback loop at the root of the symptoms, you can understand where the system might be heading. Anticipating momentum allows you to design a response calibrated not for the current state of the system but for its exponentially worse position in the future. You can budget based on expected future fires. You can plant community orchards based on expected future food insecurity. You can make sure the new clean energy grid is designed with future demand in mind.

Understanding the momentum of reinforcing feedback loops helps you anticipate the changing landscape of attitudes and support as well. If you know a problem is likely to grow exponentially, it's safe to assume that the appetite for radical solutions will also increase. This is a way of using the momentum of a dangerous reinforcing feedback loop against itself. As an escalating feedback loop leads to more and more serious consequences— such as death, illness, and economic turmoil—for people who did nothing to create it and don't benefit from it, those people, if they've been on the sidelines, may no longer sit back. For multisolvers this can mean more potential partners and more power to bring about solutions.

These consequences aren't to be celebrated, but they can be anticipated. And because every change in a system gives rise to other changes, the consequences of an out-of-control reinforcing loop can rouse other parts of a system to action. Your strategy can help ensure that rousing to action is as robust as possible.

Slowing an out-of-control feedback loop isn't easy. If you intervene early, know that you will be less alone over time. More and more people will see the problem as it grows. You may not be able to predict the moment when other parts of a system will be ready to act, but trust that reinforcements are coming, and be prepared to welcome them and channel their energy. Ideally, you'll be preparing for their arrival long before it comes.

Tapping the Power of Reinforcing Feedback

Please don't get the idea that reinforcing feedback is bad. It's essential for life. Reinforcing feedback is what brought you from a single fertilized ovum to a fully formed baby. Reinforcing feedback is how good ideas spread. It is how innovations reach those who need them. Reinforcing feedback can take stuck systems and launch them forward into new configurations. Crucially, it can help you do all of this without exhausting yourself. Reinforcing feedback loops, once launched, power their own change. This makes reinforcing feedback one of the most potent multisolving tools at your disposal.

The amount of time we have to ameliorate the many dangerous crises we face is short. We know our societies need big change, and yet our efforts often feel tiny. We find ourselves constantly outnumbered or outspent.

Reinforcing feedback offers solutions to problems like these. Like the tiny spark that sets off a wildfire, reinforcing feedback produces big results from small starts. If you have big goals and limited resources, you need strategies that tap the power of reinforcing feedback.

When change feeds on itself, ideas that begin as tiny seeds grow into mighty trees. If those initial seeds have in their DNA the power to address more than one problem at the same time, then reinforcing feedback has the power to amplify their multisolving potential.

At first the seeds might appear dormant, their growth perhaps almost unnoticeable. But if reinforcing feedback continues uninterrupted for long enough, green shoots can emerge and grow quite large—large enough to change the world.

Movements grow via reinforcing feedback. If each participant at a march brings a friend to the next one, and that friend brings a friend, who brings a friend, that is a recipe for exponential growth. Movements can build power for addressing root causes, advocating solutions that ripple out to address multiple problems at the same time.

Ideas themselves spread via reinforcing feedback. When an article "goes viral" or a new idea spreads by word of mouth, that's reinforcing feedback. If a multisolving project achieves great results in one city and people from other places learn about it and give it a try, they may well have successes too. More successes mean more experiments mean even more successes. That's reinforcing feedback!

Courage, passion, and commitment grow exponentially as people find meaning in their work or their activism. Passion and commitment lead to results, which bring a sense of accomplishment and even more passion and commitment. Courage is contagious. When you are brave, I feel braver. And when I act bravely, it may inspire others, on and on in a rising cycle of courage.

Although reinforcing feedback can lead plant and animal species to extinction, the reverse is also true: threatened species can recover via the same mechanism. If the pressures on an animal are removed, more of their progeny will survive to reproduce; with more progeny surviving lies the potential for even more "progeny of the progeny." Restoring species and ecosystems can produce multiple benefits—including, in some cases, the protection of human livelihoods, soil and water restoration, and climate change resilience. In this sense, supporting reinforcing feedback can be at the heart of a multisolving strategy.

New renewable technologies such as wind and solar power are yet another example of reinforcing feedback. For every doubling of installed capacity, the cost of these technologies falls. Installers become more efficient and manufacturers tap into economies of scale. The technologies then become more affordable to more people, leading to more installations and costs falling even further. The power of this feedback loop is significant: in the last decade the cost of large-scale solar power has fallen by more than 80%, and the cost of onshore wind power has fallen by almost 40%.[6]

Since wind and solar can replace health-harming fossil fuels while providing good jobs, they are another multisolving solution with the potential to grow exponentially.

These examples are just the tip of the iceberg. In every direction, latent reinforcing feedback loops exist that can support multisolving. They are just waiting to be encouraged along.

———————

Reinforcing feedback loops, or at least the potential for them, are latent all around us. As more people install solar panels, more people become familiar with their advantages, and more people want to install them on their own roofs. The following are a few simple ideas to keep in mind about incorporating reinforcing feedback in your multisolving efforts.

Pay attention to every link in the reinforcing feedback loops you hope to tap, no matter how many steps there are, from A to B to C to D and back again to A.

Consider a feedback loop that could boost the membership of your multisolving organization: more people attend your events, leading to more people feeling inspired by your work, leading to more people attending your events. Sounds good in principle, right?

Let's say everything starts off great. Your first event receives excellent publicity and attendance beyond what you'd expected or even hoped. It ends up being a great night that truly is inspiring. But . . .

You forget to tell people about the date and location for next month's meeting and lose the sign-in sheet so you aren't able to contact anyone. Your reinforcing loop doesn't get off the ground, for reasons that wouldn't have been too difficult to fix. All the energy you put into invitations and programs has less impact than it could have. You had a great event, but you haven't really gotten change to start feeding on itself.

To avoid this pitfall, be explicit about the reinforcing feedback loop you are trying to harness. What happens first? What will happen as a result of that? What do you need to have in place to make sure that A leads to B, which leads to even more A? Figure that out for each step around the loop, and make sure that whatever you discover ends up on your to-do list!

Focus on stickiness. Communications experts often talk about the "stickiness" of an idea. How likely is someone to hear the idea once and remember it? How compelled do they feel to repeat it and share it? The stickiness of an idea helps determine the strength of a word-of-mouth reinforcing feedback loop. If you want to take an idea to prominence quickly, it will need to be sticky.

Not all reinforcing feedback loops in our multisolving strategies are about ideas and communication of course, but the same principle applies. If you hope a new technology will spread using a reinforcing loop whereby satisfied customers tell others about the product, make sure the product both meets customers' needs exceedingly well and is interesting or exciting enough to make the customer want to share their experience with others. If you are trying to spread a skill via a train-the-trainer model, make sure the training curriculum is easy to use and memorable and its materials are accessible, reproducible, and shareable.

If a reinforcing feedback loop is part of your strategy, make sure you're giving it all the power and support it needs to spin and spin and spin, faster and faster, stronger and stronger.

Don't worry too much about the size or speed of your initial effort. Multisolving efforts often start small, which can feel daunting. How can you ever make a big enough difference? If that question sounds familiar, remind yourself about the dynamics of reinforcing feedback. Don't worry so much that only six people came to your initial meeting. Ask yourself whether you shared something that will get them excited enough to want to invite six more. Don't worry so much that your first-year sales were tiny. Look at their trajectory on a graph. Are they climbing, albeit slowly? Don't worry that only a few communicators seem to have picked up on your idea, listen instead to *how* they're talking about it. Do they understand it? Are they passionate? Do they need more coaching, or better talking points?

Remember the hockey stick–shaped graph of exponential growth? Slow initial change is exactly what you should expect from reinforcing feedback strategies. Instead of worrying about the size of your initial impact, try to assess how well the feedback loop itself is working. Is every link in the chain well supported? Does the reinforcing process continue to work a few cycles out, or has something changed? Do you need to adjust? How is your stickiness? If these factors are all attended to, your impact will grow.

Expect resistance. No reinforcing feedback in the real world exists in isolation. You'll likely find this out as soon as you start to gain some momentum. If you're building political power, some vested interest that's threatened by the power will try to slow your momentum. If your sales are rising, a competitor may show up.

In the next chapter, we will talk more about balancing feedback loops and the ways that systems resist change. For now, just know that few successful reinforcing loops go unchallenged or unresisted for long.

While designing strategies based on reinforcing feedback loops, ask yourself how the system might push back if you are successful. Could your efforts be discredited? How might your access to communication platforms be disrupted? And so on. Think again about the factors required for your reinforcing feedback loop to spin. Are all within your control? Could they be undermined?

Be aware, however, that the resistance could also come from within your own system. Are you prepared for your number of members to grow exponentially? Do you have the database and the staff capacity to be able to track and support and meet the needs of all those members? Is your production line prepared to meet orders if they double? How will you make sure that your pipeline of trainers can keep up with demand?

In the early days of your strategy, when change seems so slow and gradual, take some precious time to think ahead. When things take off, what might happen next?

? Questions for Reflection

- In the systems to which you'd like to apply multisolving, are any problems growing in such a way that change feeds back to create more change in the same direction? See if you can identify the reinforcing feedback loop.
- See if you can name some approaches that could slow or stop the reinforcing feedback loops you identified in your answer to the first question.
- Are you involved in any strategies that operate via reinforcing feedback? (Examples to prime your thinking: word of mouth, building community, building power, spreading a solution.)
- See if you can list the steps in the feedback loop you identified above. (Note: you should be able to show how an initial change feeds back to create more change in the same direction.)

You Bring Me Back

Every time I get off course
you bring me back.
Gently,
you walk me down
from the peaks,
with their ferocious winds
and heart-piercing views.
Again, and again, and again,
you push me upward
from the stagnant lowlands,
the deep swamps.
The highs are high,
the lows, so low.
But, there you are,
anchoring the center.
A calm, still place.
A steady goal.
A through line that wavers,
but goes on,
and on.

CHAPTER FIVE

Balancing Feedback

In nature and in other complex systems like societies and economies, patterns of boom are followed by patterns of balance. Aphids are all over the lettuce in the garden and then subside when the ladybugs move in. Fads rise and then fall away. Periods of rapid economic growth often are followed by periods of stagnation. Empires grow and decline. Tempering the explosive power of reinforcing feedback is the homeostatic power of balancing feedback.

Recognizing Balancing Feedback Loops

Countless times a day, in countless ways, balancing feedback helps you move through the world. Inside your body, balancing feedback keeps your temperature constant. It tells you to drink when you are thirsty and eat when you are hungry. It regulates your blood sugar and your oxygen levels. It helps you steer along the curving road on your drive to work. It helps you decide to pass one assignment on to a coworker so that you have the capacity to get another job done. It's there when you cast your vote in an election and when you decide your closet is too full and you give some sweaters away. Balancing feedback brings systems back to a goal or set point after a disturbance or a disruption. Let's see how it works.

Your body requires a certain ideal level of hydration to maintain fluid balance. Structures in your brain sense a rise in the concentration of sodium and other solutes in your blood. That sensation is what we call thirst. The less hydrated you are, the thirstier you feel. Thirst creates a goal-seeking behavior, which is a hallmark of balancing feedback loops—in this example, you search out a water fountain or cup of tea. As the water passes into your bloodstream, the sensing mechanism notes that, for now, your hydration goal has been satisfied. Your thirst diminishes. A change in one direction—decreased hydration—feeds back around with a change in the opposite direction, increased hydration.

Do you see how balancing feedback is different from reinforcing feedback? In reinforcing feedback loops, a change in one direction creates more change in the *same* direction. In balancing feedback, a change in one direction creates a compensating change in the *opposite* direction. While reinforcing feedback causes variables to grow or fall, balancing feedback holds them steady at or close to a set point. Because of balancing feedback's role in health, balance, and restoration, many multisolving strategies involve strengthening these loops.

When part of a system is not working as the people within it desire—food insecurity is rising in a community, for example—balancing feedback loops swing into action to bring the system closer to the desired state, in this case mobilizing to feed the community.

If bringing part of a system closer to a desired state solves multiple problems, then strengthening a balancing feedback loop is a multisolving strategy. Some strategies to reduce food insecurity feed people while also solving other problems. My colleagues and I learned about one such example in Spain: Espigoladors, an organization that focuses on making sure less-than-perfect fruits and vegetables aren't thrown away. The organization clearly describes a vision of multisolving on its website: "We are a social enterprise model which works for three social needs at once and interconnects them: fighting for a better food usage, guaranteeing the right to a healthy diet and creating job opportunities for collectives at risk of social exclusion."[1]

The people most at risk for hunger are invited to participate in gleaning and processing what would otherwise be wasted food, gaining skills and employment in the process and being welcomed into a community. "In Espigoladors we give second opportunities to ugly and imperfect fruits and vegetables and to beautiful people."[2] Since food waste is an important contributor to climate change, the work of Espigoladors multisolves in yet another way!

Not all balancing loops display this level of multisolving. The same goal of reducing the level of food insecurity could be met with a different balancing feedback loop, perhaps providing donations of food to those in need. While this is an admirable and potentially lifesaving balancing loop, it doesn't offer the other multisolving benefits of reduced food waste, job training, community building, or climate protection.

Sometimes powerful forces within systems are invested in a different idea of what the status quo of the system should be. Their desired state is "just as it is right now." This is a second reason it is wise to be aware of balancing feedback. Status quo goals might include a certain level of profits, a certain level of political power, or media ownership for a minority of a population. From their point of view, your multisolving efforts are moving the system away from a goal, a goal that probably has balancing loops in place to undo the "disruption" that results from your efforts.

Knowing that many multisolving solutions strengthen balancing feedback loops, and knowing that the successes of multisolving may create the need to cope with countervailing balancing feedback, let's look more deeply at how to recognize and work with balancing feedback.

———————

If reinforcing feedback in a system feels like a runaway train, what does balancing feedback feel like? Steadiness, recovery, a return to normal following disruption. Balancing feedback loops carry us along in waves that crest and fall. Things rise above a set point, fall to meet it, fall below, and then rise again. Balancing loops bring healing. When you stand up and stretch after sitting for a long time, that's a balancing loop. That feeling you get when you return to work after a long vacation? That's a balancing loop. So is that vacation feeling, a week of sleeping late and lying around in the sun after a long, hard push at work. Balancing loops interrupt movement when it's gone on too long in any one direction.

So that's what balancing feedback loops *feel* like. What would one look like on a graph? Imagine a graph charting a room's temperature over time, where the system's goal—the thermostat's setting—is 65°F. A mark on the graph at 65° would represent this goal as the starting point. On a cold morning, the room might cool off. Let's say the temperature falls to 63°. Put a little mark on the graph at 63°. What happens next? The thermostat recognizes that the temperature of the room is below the goal, and so it delivers heat. The temperature then rises to meet the goal of 65° and the furnace turns off. Make another mark of 65°. Maybe it's a sunny day, the room warms up to 70°, and the thermostat notes that now the room is warmer than the goal. Make a mark at 70°. The air-conditioning system turns on, and the room falls back down to 65°, so make a mark on the graph there. For the rest of the day and the next day and the next day, unless the thermostat is adjusted, that's what the temperature in the room will do: rise and fall just a little bit above and below 65°. If you connect all of your points with a line, it will resemble something like a wave, a pattern of oscillation around a goal; that's the classic graphical behavior of balancing feedback loops.

Balancing feedback loops, including the thermostat example, are made up of three parts: a goal, a gap, and a way to close the gap.

The *goal* is the desired level of a certain variable in the system, for example, a room temperature of 65°F or your body's optimal blood sugar level.

The *gap* is the difference between that goal and the variable's actual level: the difference between the temperature of the room and the desired temperature of the room or the difference between your actual blood sugar and your optimal blood sugar.

A *way to close the gap* is specific to the system. When the thermostat detects a gap between the room temperature and the desired temperature, it activates the furnace. When your body detects too much glucose in your bloodstream, your pancreas releases insulin.

When action is taken to close the gap, the actual state of the system moves nearer to the goal until the gap is no more. At that point the gap-closing action ceases until a new gap is registered and the gap-closing action is once again called back into play.

Stock levels are often regulated by balancing loops. The goal is a target value for the level of the stock. The gap is the difference between the actual stock level and the goal. The gap feeds back to regulate the inflow to the stock or the outflow from the stock, or both. For instance, a furnace controls the flow of heat into the stock room temperature. Air-conditioning regulates the flow of heat out of the room.

With the goal/gap/action structure of balancing loops in mind, you might now recognize how many overlapping crises we're facing result from ineffective balancing loops. Perhaps the system isn't registering a gap. Maybe there's a lack of consensus about a goal. Maybe the gap-closing mechanism is not strong enough.

The biodiversity crisis is the gap between the desired health of Earth's ecosystems and their actual health. Habitat conservation, pesticide bans, and land restoration are efforts that work to close the gap. But at least for the moment, the pressures on biodiversity are growing faster than the gap-closing mechanisms we have in place.

The climate crisis is the gap between the desired concentration of CO_2 in the atmosphere and the actual concentration. Policies to prevent extraction of fossil fuels, spur investments in infrastructure for clean energy, institute carbon pollution fees, and draw carbon back down into soils and forests could close the gap. The climate movement pushes and pushes to strengthen these balancing feedback loops. But as the deepening climate crisis shows, so far these balancing loops are too weak relative to the drivers of climate change.

A city's homelessness crisis is the gap between the number of homeless people and the city's goal of housing for all. When the city incentivizes affordable housing or connects homeless people to services and support, it is activating balancing feedback loops. If the crisis persists, it means those balancing loops are too weak relative to the forces that drive homelessness.

Balancing feedback loops are often about "enough." They sense whether you have enough food in your belly, money in your bank account, or status in your department. When a feedback loop senses that a goal is met, it halts the gap-closing action. But balancing loops are also about "ouch." When you place your finger too close to a hot flame, a balancing feedback loop jerks your finger back. When you ask your teenager to turn down the music that is making your head pound, that's a balancing feedback loop. A balancing loop is at play when a society's grief at the sight of a clear-cut forest leads it to enact environmental protections.

The culture that I come from—Western, industrial, growth-oriented culture—errs on the side of overlooking balancing feedback loops. We override them. We break them. We refuse to listen to their signals. In fact, we have entire institutions and industries dedicated to the explicit goal of interfering with balancing feedback. The advertising industry, for one, works so hard to convince us that we can never have enough or be enough. Having lost the ability to sense enoughness, we are disconnected from essential balancing feedback. This is not an accident. In fact, our economic system depends on the suppression of some of our ability to sense enoughness.

The dominant culture also has trouble heeding the feedback loop that says "ouch." Look around at our world, with its ecological devastation and human suffering. Where are the feedback loops that sense harm and enact a response? These too are not missing by accident. They are suppressed by political and economic systems.

Systems like white supremacy, patriarchy, and extractive economics all undermine balancing feedback loops. When the dominant culture asserts that some people matter less than others, based on gender or race or other factors, that culture is saying, "feel free to disregard signs of pain or distress from those people." In other words, *feel free to disregard balancing feedback.*

Likewise, when a culture asserts that Earth is a collection of resources, rather than a web of fellow beings, it suppresses a balancing feedback loop. No need to stop mining, harvesting, or dumping. No need to feel sadness, anger, or fear at environmental degradation. No need to respond to the balancing loop that is the rainforests, the oceans, or the tundra screaming "ouch!"

Systems can change when a critical mass of people open themselves to balancing feedback and organize to do something about it. I see you, worker sewing T-shirts in Bangladesh, and I will shift my spending to a worker-owned cooperative. I see you, young Black man terrorized by racialized state violence, and I will change how I vote and where I donate and start knocking on doors. When we open ourselves up to balancing feedback, we open the possibility of transforming systems.

Journalists, artists, activists, and scientists play an important role when it comes to balancing feedback. Often, they reveal the gaps between where we are and where we say we want to be. A scientist might measure lead levels in a community's drinking water. A journalist might break the story that exposes the inequities causing the water to be contaminated in certain zip codes but not others. An artist might film a documentary showing the impacts in a way that touches people's hearts. Activists, with their marches and petitions and fiery speeches, highlight injustices, inequities, and environmental harms. They shine a spotlight on the places in systems where the goals or the efforts to close gaps are insufficient.

Simple awareness of a gap doesn't change systems on its own. A whole society can know that something is not right or that danger is looming and still not be organized in a way that enables corrective action. That signal of "ouch" isn't always enough to spark change, but without the "ouch," the odds of change are even lower.

When you inquire as to the well-being of another person, you are opening yourself to balancing feedback. The same is true when you ask someone for consent. *Is what I'm doing causing harm? If I were to do it differently would that be better? Would it help you? What do you need?* Individuals can ask these questions, but so can governments, institutions, and corporations.

Opportunities for balancing feedback sit latent in our systems much of the time. The potential is there for us to understand the impacts of our actions, but sometimes we move too fast and forget to ask. Indoctrinated by various systems of superiority, we often think we don't need to ask. Sometimes drawing on balancing feedback is as simple as asking a question and listening carefully for the reply.

Tapping into balancing feedback can transform the world. From the Civil Rights Movement to the environmental movement enlivened by Rachel Carson's *Silent Spring*, when large numbers of people decide that a gap between the state of a system and a desired state is intolerable, the world shifts.

Sometimes balancing feedback is lost in the complexity of systems, and sometimes it is purposefully buried or distorted. Bringing better, clearer, more accurate information about the difference between a goal and the actual state of a system can be an excellent way to multisolve.

Strengthening Balancing Feedback Loops

There are several potential places to intervene in a balancing feedback loop: the goal, the gap calculator, and the gap-closing mechanism. For a balancing feedback loop to work well, each of these elements must be strong and they must be strongly connected. Here are some important considerations.

Make sure the goal is strong and clear. Without a goal, there is no gap. And without a gap, there is no reason to act, nothing to spur corrective action to change the condition of the system. It makes sense then that many of our most passionate political battles are about the goals of balancing loops. Who matters in the system? Is our goal to reduce the mortality rate of pregnant Black women to as close to zero as possible? Or are we willing to tolerate inequity in pregnancy outcomes? Is our goal to eliminate lead in the drinking water supply? Is our goal zero increase of carbon dioxide in the atmosphere?

When thinking about the goals of balancing loops, make sure they are goals in practice and not just in theory. A goal in practice drives a gap-closing change process; a goal in theory is something people talk about, but it is unconnected to anything with the power to close a gap.

Changing the goal of a balancing feedback loop can be a transformative action within systems. Some of the most important social and political movements in our history have changed goals. In the United States, the Clean Air Act set standards for certain pollutants in the air.[3] The Sustainable Development Goals championed by the United Nations offer the world seventeen clear goals for the well-being of people and the sustainability of economies.[4]

Goals matter at smaller scales too, and goals have a role to play in multisolving. One project my colleagues and I learned about was a collaboration between Barts Health NHS Trust and Global Action Plan in London. Operation TLC set a goal of turning off all equipment in the hospital when it was not in use.[5] This energy efficiency goal created a training program for hospital staff and a system to show at a glance which equipment could safely be turned off. Not only did the program help the hospital reach its climate goals ahead of time, it also improved patients' comfort and ability to rest and recover because the hospital became a quieter place.

Today many goals within our systems are strongly contested. What is the goal for the level of CO_2 in the atmosphere that we will work to achieve? What is the goal for the percentage of the population with access to affordable health care? The goals we choose today will influence how well (or how poorly) we multisolve and how well we navigate future challenges.

Discern which goals can drift and which to hold fast to. Some losses—including climate change impacts, economic instability, ecosystem disruption, and more—seem unavoidable given the pressures on natural and human systems at this point. If we are trying to keep everything "normal" in the face of these impacts, then we will be directing more and more of our energy into more and more balancing feedback loops. Each destabilizing event will create a gap between how the world is now and how it was. We could put all of our wealth and ingenuity into closing those gaps and trying to preserve a sense of normal. But as deeper and deeper pressures destabilize systems, it will take more and more effort just to stay in place, never mind what is required to face and avert mounting crises.

In a changing world, sometimes what is required is to allow some of your goals and expectations to shift, like how many times a week you need to eat meat, how many trips you need to take in a single-occupancy vehicle, or how much "productivity" your team at work can accomplish in the face of multiple stresses on their communities. With only so much time and energy, we need to be discerning: Which balancing feedback loops are essential, and which are less so? Trying to support the lifestyles we've become accustomed to even as supply chains become brittle is a balancing loop. Trying to keep global CO_2 levels in a safe range is also a balancing loop. I'd argue the second is more fundamental than the first. And too much effort on the first might actually make the second harder to achieve.

Which of our needs are core needs, essential to well-being? These are goals worth fighting for. Others may not be. What can we give up? What losses can we accept, and which ones will never be okay with us? Only people and communities in conversation with each other and aligned with their highest values can answer these questions, but tumultuous times demand that we ask them.

Make sure there's a mechanism to calculate a gap. For a balancing loop to be effective, you need a way to measure the state of the system, and you need a way to compare that measurement to the goal. A political leader might declare the goal of ensuring every child receives adequate nutrition. But without a budget to measure how many children are receiving adequate nutrition, there's no way to know the size of the gap. Without an identified gap, a balancing loop is weakened; it's harder to drive action.

I learned this in a big way around the time of the Copenhagen round of global climate change negotiations. My colleagues and I were fairly new to these high-level talks. We knew that there was a goal: to limit global temperature increase to 2°C. We knew countries were pledging actions to reduce their greenhouse gas emissions by a certain amount and by a certain year relative to a base year. Countries were using different target years, different base years, and different definitions of "reduction." With all these diverse pledges, the United Nations couldn't calculate how close to the goal the pledges would take the world. They couldn't calculate a gap. Luckily, around that time technical breakthroughs by our group and others made this gap calculation possible. In fact, every year since, there's been an annual Emissions Gap report led by the United Nations Environment Programme.[6] The findings of the report help global leaders and civil society understand how much more climate action is still required.

Gaps, like goals, are another point of contention within systems. If a balancing feedback loop works against the perceived interests of some stakeholders, they may act to hide information about the gap. Starved for awareness of a gap, a feedback loop is disabled.

Information about gaps needs to be clear, accurate, and timely. If there is a long delay between measuring the gap and taking action to close it, the balancing loop will forever be responding to outdated information. If the gap has grown bigger while the gap-closing mechanism sluggishly responds, then the action will be too small to be effective in restoring balance.

When fossil fuel interests fund disinformation campaigns, they are trying to suppress information about the gap between safe levels of greenhouse gases and today's levels. They are trying to delay the activation of a balancing feedback loop that would reduce fossil fuel use.

As someone working to steer systems toward multisolving outcomes, you need to ensure that the right information is being measured. You need to fight to ensure it will be used in a timely fashion to calculate a gap. Campaigns and social movements do this all the time, though they may not talk in terms of balancing feedback loops. They draw attention to a gap. They force us to look at the ways in which our systems are not living up to our goals for them.

Sometimes leaders expose information about the state of the system, despite the efforts of others to suppress that information. Whistleblowers document unsafe workplace conditions and dangerous levels of environmental toxins. Such interventions shine light on the actual state of the system. They make us see the gap between where we are and where we want to be. In doing so, they can activate balancing feedback loops.

Invest in the ability to close the gap. It's important to set goals for the levels of key stocks within the system. It's critical to know when there is a gap between that goal and its actual level. But knowing that a gap exists is not enough to change the behavior of the system without a way to change the gap.

All balancing feedback loops employ some mechanism to change the state of the system. They somehow actuate a change to the inflow or outflow affecting the stock of concern. Sometimes the action is quite direct; sometimes it involves a long and convoluted chain of causation. But without it, a balancing feedback loop is not a closed loop at all.

Think back to chapter 3, where we discussed the many ways rates of flows can change with interventions like investment, policy, and planning. Those interventions work because they kick balancing loops into action. Investment in education tries to close a gap between present and desired educational outcomes. Environmental policy tries to close a gap between people's goals for air and water quality and their actual quality. Votes for parks close a gap between the actual and desired amounts of green space in a city.

Sometimes a goal may be clear and shared among stakeholders and the gap may be obvious, but the missing link to steer the system is the capacity to act. When this is the case, improving the systems may require building the power sufficient to activate a strong balancing loop. Think back to the previous chapter on reinforcing feedback, and you'll realize that sometimes, to activate a balancing loop, you'll need a reinforcing one first.

Disrupting Balancing Feedback Loops

If your efforts to multisolve feel stuck or ineffective, it may be that a balancing loop is in the way. There is no simple formula for navigating this sort of resistance, but it helps to know what to look for.

Think hard about the key stocks whose levels you are trying to change. Identify the inflows and the outflows. Think about the other forces within the system. Are there any entities that would prefer to keep that key stock at its current level? Might some organization or individual be trying to move the stock level in the opposite direction of your efforts?

If your multisolving is successful, it may create gaps elsewhere in the system, changing a stock or a flow rate. Another stakeholder may register that change as a deviation from *their* goal. They will sense a gap, and their part of the system will, if it can, try to close *that* gap.

Learning to expect and counteract or disable such balancing pushback is part of the art of multisolving.

Anticipate the consequences of your success. All else being equal, the bigger the change you create in a system, the bigger the pushback will likely be. Part of planning for and envisioning success should involve answering at least two questions:

- Suppose we succeed beyond our wildest dreams, what balancing feedback loops might that success trigger?
- How can we be prepared to diffuse those loops?

Ask these questions early and often. You should be preparing for systemic pushback long before you trigger it.

You won't always be able to predict what balancing feedback loops will be triggered. But whenever you can, getting ahead of them can help make sure your interventions within systems are as effective as possible.

Try to turn competing goals into synergistic ones. Sometimes both you and another force in the system will want to see levels of the same stock move in opposite directions. You'll face a struggle over your opposing goals. At other times you might experience pushback because a change you've made has rippled out to affect some other part of a system that matters to someone else. In these situations, there is good news and room for creativity, especially if you bring a multisolving lens.

Here's an example. Imagine a group of parents successfully lobbies the school board to enact a 10% increase in music spending for the next school budget, which will be paid for by an increase in the property tax rate. However, some people in town—mainly senior citizens living on fixed incomes—are alarmed. If their tax bill goes up, they might have trouble paying other bills. The new proposal triggers the seniors to act: letters to the editor of the local paper, a recall vote directed at the school board. Suddenly the music program feels at risk. Two balancing feedback loops are operating at the same time. One group is trying to close a gap in desired music instruction. The other is trying to close a gap in their ability to pay their bills.

The parents' group could double down. They could write their own op-eds on the importance of music to educational outcomes. They could recruit other families to their organization. They could put forth their own candidates for the upcoming school board election. The ultimate result will be determined by the relative strength of the two balancing feedback loops. There would in this case be a winner and a loser but perhaps only temporarily. If the balance of power shifts in the next budget cycle, the outcome could shift too.

The parents' group has another option though, one that might seem more challenging at first but could have a more lasting and satisfying outcome. Is there a way to close both loops together rather than have them fight one another? Could there be a property tax rebate for senior citizens? Why are seniors in such duress in the first place? Surely a town could have thriving children and thriving elders? What about a different financing mechanism that would not impact taxes? Could the music program be staffed by volunteers?

If your efforts are experiencing pushback, it is worth trying hard to understand its sources. At least some of the time, there may be creative ways to meet both goals at once. That can be an efficient way to disable a balancing feedback loop standing in your way. It can even lead you to solving multiple problems.

Build the power to resist pushback. Sometimes, of course, it's not possible to meet multiple goals simultaneously. If my goal is zero carbon emissions and yours is continued profits from burning of coal oil and gas, we are going to be struggling against each other. There will be opposing balancing feedback loops.

If your goal is racial equity and your governor is a white supremacist, there is no mutually satisfying "middle ground." You are focused on balancing loops with different goals for the state of the system.

In such cases, the behavior of the system overall will depend on the relative strength of the opposing balancing feedback loops. Is your capacity to close the gap motivating you stronger than someone else's ability to close the gap motivating them? This of course is a matter of power. Do you have the power to overcome a balancing feedback loop that is keeping you from reaching a goal?

In a standoff between balancing feedback loops, your mind should turn to reinforcing feedback. If you need allies, money, attention, numbers, colleagues, or influence to resist pushback, then you need a process where change builds quickly on itself.

More Than the Sum of the Parts

We've now completed our tour of the elements of systems: stocks, flows, reinforcing feedback loops, and balancing feedback loops. By looking at these parts of systems one by one, we have started to see hints of the complexity that arises from whole systems.

In real systems, in whole systems, many feedback loops all spin at the same time. They interact and intersect. Some of them spin quickly. Some move very slowly. Reinforcing feedback loops push the system toward explosion. Balancing feedback loops try to hold it steady.

All of this can quickly overwhelm the human mind. How can we keep track of each feedback loop and the ripples of change they create? How do we understand their interconnections, their relative strengths, and their relative speeds?

The short answer is, not very well—at least not for the dominant cultural, political, and economic interests and institutions on Earth right now. Too often, these institutions squash feedback and ignore connections. They tap into reinforcing feedback loops with casual disregard for their explosive power. In their ignorance of how systems work, they create harm and catapult us toward catastrophe.

The good news is it doesn't have to be this way. While complex systems are impossible for us to fully predict or control, we can work with them. We can steer them. We can develop codes of conduct that are both practical and ethical for a world of complexity and interconnection.

? Questions for Reflection

- Where do you see effective balancing feedback operating in your life or your work? What desired goals are being met? What mechanisms are strong enough to bring your system back to balance when it gets destabilized?
- What crises in your world might be resulting from balancing feedback loops that are too weak to close a gap? See if you can identify all the parts of that feedback loop: the goal, the way to measure the gap, and the way to close the gap. Which parts aren't working as needed?
- What part of your work/life feels stuck? See if you can identify any balancing feedback loops that might be responsible. Do you have any options for weakening those feedback loops?
- If each of the following balancing feedback loops were stronger or more effective, how many different benefits might arise? See how many you can list.
  - Creating forest reserves to address declining hectares of global forest
  - Responding to declining hectares of global forest by returning land to Indigenous stewardship
  - Building a new stock of affordable housing to combat a city's housing crisis
  - Increasing a city's amount of green space in response to an increase in the number of high heat days per year
  - Building cooling centers powered by renewable energy in response to an increase in the number of high heat days per year

Tangled

Are you gigantic?
As big as the Earth?
Or actually, if I may digress,
since it is the light of a star
    pumping your heart
(after its journey of light and energy
through croplands and beetles
    and honeybees to become your
    breakfast, pumping your heart).
Since it is the light of a star,
    animating your eyes/
    brain/mind/heart
interpreting these small, strange,
    black marks on white paper,
paper that was once tree
    animated also by starlight,
that was once pulp, then
    paper, then book . . .
Oh, but here I digress again,
forgive me, my thoughts are tangled,
like the world.
My thoughts,
animated by starlight and captured
    by these small, strange,
    black marks before you
(I wonder, are these words
    frozen starlight?)
But I digress . . .
What I mean to ask is this:
Since it is the light of a star
    that animates you,
are you as big as the solar system?

Or are you tiny?
A tiny speck of attention.
For an instant are you as small
    as the tickle of that rough tag
    on your new shirt,
that annoying, infuriating tickle
    on the back of your neck?
The only tickle on the only
    neck in the universe that
    is exactly just like that?
Or are you at once tiny
    AND gigantic.
Tigantic—if only that was a word.
At once as old as the birth of
    the sun and brand new,
remade just this morning
    at breakfast?
Is the whole universe
    swirling through you
while you decide to reach up and
    scratch that troublesome spot
    on the back of your neck?
I know, it is all ridiculous, these
    digressions and tangles
when you have bills to pay and
    a pile of reports on your desk.
I know.
I myself, though you may
    not have noticed it, can
    get lost in the tangle.
But really, can you find any flaw
    at all in what I have to say?

CHAPTER SIX

The Behavior of Whole Systems

The smoke alarm was blaring. Steam swirled around me; water pooled at my feet. I had caused a "boilover" in our community heating system.

My neighborhood of twenty homes is heated by a shared wood-burning furnace. My neighbors and I take turns loading wood into the firebox, which warms up a tank full of circulating water that in turn carries heat to our homes, including our hot water tanks.

The goal when managing the heating system is to keep the water as warm as possible without it reaching the boiling point. If you add too much wood for the given conditions, you end up with the dreaded boilover, which is just what it sounds like. When the water reaches the boiling point, the steam needs to go somewhere, so steam and boiling hot water come burbling out onto the floor of the building that houses the furnace. There's no fire danger, just a lot of steam in the air that fools the smoke detector. Still, boilovers make a mess and require extra work to refill the tanks to make up for the water that has sloshed out of them. Nobody wants to be the one who causes the boilover. So, what went wrong? Like many of my neighbors before and since, I was fooled by the dynamics of a complex system.

In this scenario, the amount of heat in the water tank is a stock. The stock of heat is filled by energy transfer from the burning logs. The stock of heat is drained whenever the thermostats of the houses attached to the system register a temperature below a set point. The thermostats "call" for heat, and hot water from the furnace loop runs through radiators, delivering heat into the homes.

When demand for heat is high, boilovers are unlikely. Heat flows to the houses as fast as you can add it by adding logs to the firebox. At six o'clock on a dark, cold morning, when many of my neighbors are turning up the thermostat and running hot water, there's high demand. The colder it is outside, the more quickly homes cool down, increasing the neighborhood's demand on the heating loop.

Clearly there are a lot of flows in this system: from wood to the heating loop, from the heating loop to the houses, and from the houses to the outside air. But there are still more. Our homes are passive solar, capturing a flow of solar energy when the sun is shining. That represents a second inflow of heat, alongside the flow from the heating loop. All else being equal, it will offset the need for heat from the wood-burning part of the system.

There's feedback in the system too, with the thermostat regulating a balancing feedback loop inside the homes. Another feedback loop is engaged every time one of us decides how many new logs to add to the fire. We want to add enough logs to make sure there's heat for all the homes but not so many as to cause a boilover. So we have many flows and feedback loops, including several in each home (thermostats with different set points for water tanks and space heat and perhaps floors set to different temperatures).

Our little system also has delays. Once you add new logs to the fire, they burn for around two hours. To manage the system skillfully you can't just think about the current state of the system when you add logs. You have to anticipate how the state of the system might change during the lifetime of that load of logs.

Now you can understand how I caused the boilover.

I added plenty of logs for the cold, dark morning, but within an hour the sun came out, strongly. All the houses warmed up quickly. Families headed off for school and work. Demand for heat fell and with it the outflow of heat to homes. But the outflow of heat from burning logs into the heating loop? It was still going strong. Thus, the boilover.

Ours is a simple system with a handful of flows and a small number of feedback loops, and around forty adults who have to coordinate to manage it—child's play when compared to the economy of a nation or the dynamics of the climate. But still, understanding each flow and feedback loop in isolation isn't always enough. Understanding the behavior of parts isn't always enough to understand the behavior of wholes.

Even with a system as simple as ours it's easy to get fooled. And of course, we depend on systems much more complex than this one every day, from complex ecosystems to citywide energy systems to national food systems.

The surprises that emerge from systems aren't always so unwelcome.

I once lost touch with my organization's project partners for a couple months during a pause in our shared work in a United States city. When I eventually checked in, there was exciting news. To our surprise, the city government had released a new plan that included goals on infrastructure equity that were closely aligned with our own. The language of the city's plan was remarkably like the language our network had been honing for more than a year. Many of us, veterans of campaigns that went on for years with-out ever getting language or policy points officially adopted, felt stunned. How had it happened?

What we learned was that, as the plan moved through the city's policy process, one staffer—who had been a part of our network discussions—added the language into the plan. The language was then reviewed and approved higher up in the department. It was a simple as that, and almost effortless from the perspective of our project. But was it?

For several years we'd been convening a group that included the city official in question, as well as members of the community living with the effects of the city's decisions. We invested in facilitation and reflective conversation. Relationships slowly strengthened. We built joint maps of important elements of the system that was the city, what was working and what could be improved. As a result of all of these activities, we changed how the parts of the system connected to one another. Out of that changed set of connections came a shift in the system that none of us predicted.

Around fifty people were involved in the project, each of whom was also part of additional organizations and communities that accounted for hundreds more. It wouldn't even be possible to list all the flows and feedbacks in this system. Flows of ideas and influence. Flows of changed perspective. Flows of money and collaboration and new partnerships. Feedback loops of learning that challenged assumptions. Feedback loops of sharing power. We couldn't control all that complexity. At best we facilitated and nurtured it. Everyone involved supported the system in evolving toward our shared goals.

We created the conditions for the system to surprise us.

———————

Within the systems to which we belong, we often focus our attention on specific goals and outcomes. If we are trying to multisolve, we may be holding several goals simultaneously. We'd like our company to be more profitable or our community to feel safer. We'd like to increase the amount of carbon in the soil or eliminate child hunger in our city. Eyes on the goal, we act. Sometimes we are pleasantly surprised. Sometimes we get a very different result from what we hoped—swirling steam, smoke alarms, and water on the floor. Systems often seem to defeat our impulse to manage and control them.

The consequences—whether to a government budget or an endangered ecosystem—of being surprised by a system can be terrible indeed. The price of unmanageability can be existential crises. And too often those crises affect people who had no power or influence on the catastrophic decision in the first place. Facing converging crises, it is easy for us to want a sense of control. When you're committed to multiple critical goals, from health to justice to sustainability, it's understandable to want to bend systems to your will. And yet the nature of complex systems is such that they often react counterintuitively, evading our attempts at control.

What, then, are we supposed to do in the face of crises and rising danger? Well, it turns out . . . quite a lot.

We won't find our way through tumultuous times by seeking control and fulfillment of all our predictions. We won't have much success solving multiple problems at the same time that way either. In fact, the quest for control may have contributed to some of the problems we experience as tumultuousness in the first place.

There is a way, however, of approaching change that could help us through crises while taking advantage of opportunities to create deep and lasting change: working *with* systems instead of trying to control them. This way of working—and being—could help us get to the heart of what is generating converging crises in the first place. To understand the potential, we first have to understand more about the behavior of whole systems, why that behavior sometimes surprises us, and why our interventions don't always produce the results we want.

Whole System Behavior

Why do some systems hum along steadily while others produce chaos and destruction? Why do some appear to work so well for so long before surprising everyone by imploding, while yet others resist all efforts at improvement for so long, only for the tiniest of changes to unlock positive transformation? And how can we both recognize and seize moments of opportunity to shift the structure of systems and steer them in ways that solve multiple problems at the same time?

These questions rarely have easy answers. Systems behave differently for many reasons, all operating and influencing one another simultaneously. Systems do what they do because of their history. They do what they do because of their interconnection. They do what they do because of nearby influences and faraway influences. In other words, a system's behavior emerges from its structure, as shown in the top two layers of figure 6-1.

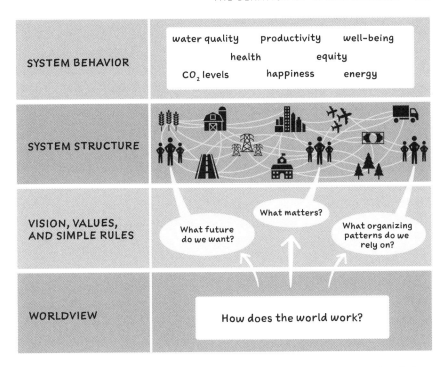

Figure 6-1. The behavior of a system emerges from the system's structure, which includes the physical elements of the system, its rules and incentives, and the flows of material and information through the system. The system structure is shaped by the visions, values, and simple rules embodied by the people participating in that system. In turn, the visions, values, and simple rules are shaped by worldviews, people's beliefs and assumptions about how the world works.   *Illustration: Molly Schafer.*

A system's structure includes both physical and intangible elements. Cities, for example, are systems that include stocks like buildings, roads, schools, and businesses and flows such as energy, waste, and natural resources. Cities also include the people and the choices they make, as well as the laws, rules, economic incentives, and customs that influence those choices. In addition, there are interconnections between the many parts—all the flows of money and information and the feedbacks that keep some parts of the system in balance and cause others to grow. The feedback loops, the thresholds, the goals and gap-closing mechanisms of balancing loops, the stocks, and flows, all are included in the systemic structure.

The structure creates the behavior, and sometimes the behavior is not what we expect, which is why it helps to pay attention to a system's structure.

In this section we will do just that. We'll go through some of the ways that systems behave that tend to surprise us and dig into how their structures generate those sometimes-surprising behaviors. We will look at these structural causes one at a time, but as you read along keep in mind that at any one moment, for any given system, several of these principles will probably be in operation at the same time. Surprises on top of surprises combined with more surprises—systems surely keep life interesting.

Even the simple elements of systems can fool us. Even systems with a small number of elements can behave in counterintuitive ways. As we introduced stocks, flows, and feedback loops we named some of those counterintuitive behaviors, so I won't repeat them in detail here. But never forget: stocks, flows, and feedback each have the ability to surprise us, even on their own. If a stock that behaves in surprising ways influences a flow or a feedback loop, then that flow or feedback might behave in a surprising way in turn.

Many interacting parts make for surprises. Systems with only a few elements can surprise us with their behavior. What happens then when a large number of elements are connected together?

A single stock can have many inflows and outflows. Imagine the manager of a busy shopping center tasked with enforcing the fire code's occupancy limit. The center has four entrances and six exits, each located on different floors, out of sight from one another. People are constantly entering and exiting, pushing past each other. Some walk in only to turn right around again to leave. Others linger and browse and sit in the corner with a coffee and a pastry. The manager frets. Is the building above the occupancy limit? How can the manager regulate the situation with all the comings and goings? And that's one stock and a few flows, without any feedback loops.

The amount of carbon dioxide in the atmosphere is a giant stock. Think of all the flows into it! Every tailpipe on every car, every furnace and cook stove. Fleets of planes. Heating plants. And the outflows! Uptake into the oceans, into plants, and into soils. The carbon cycle also includes lots of feedback loops, of course. Reinforcing feedback drives the growth of populations and the economy and increases many of the inflows of CO_2 to the atmosphere. Balancing feedback loops emerge in response to the threat. Scientists sound the alarm. Social movements push for a transition to low-carbon technologies.

Since each part of a system has the potential to change at any given time, the interacting components can become too numerous for us to track. And even if we could somehow track all the parts and how they are changing, it would be tricky to add up the net effects. Is the store gaining or losing occupants? How fast is CO_2 accumulating in the atmosphere? How might that rate change under a policy directive to plant more trees? What if all of a country's vehicles became 10% more efficient, while the vehicles in another country became 3% less efficient?

When feedback loops change in strength, system behavior shifts. A strong balancing loop can keep a reinforcing loop in control. A weak balancing loop will be dominated by a strong reinforcing loop, and the system to which it belongs might spin out of control. That's how only two feedback loops might interact. Now imagine six or even twenty. Complex systems feature many feedback loops interacting at the same time.

The connections between a system's parts determine its behavior. But it's not just which parts are connected to each other, it's also the strength of those connections. Picture two systems, each with the same causal connections. The links between the parts are the same, but what if the connections in one system are stronger than in the other?

Democracy can flip to autocracy if enough balancing feedback loops become weakened. Take the feedback loops that together make up what we call "the free press." The strength of those loops can fluctuate. Media consolidates. Local papers decline. Social media and the profitability of click-based advertising models change cultural behavior and financial incentives. Each of these changes has the potential to weaken a balancing feedback loop in the system.

Other loops might be growing stronger. Is more "dark money" now permissible in campaign finance rules? That's a strong reinforcing feedback loop. More political influence allows some entities to profit more, who in turn can invest some of those higher profits in the quest for even more influence. Is one political party redesigning voting districts to further entrench their power? That's another reinforcing feedback loop. The power to define voting districts influences who wins elections. And who wins elections determines who has the power to further define voting districts.

The outcome for democracy will depend on the relative strength of all of these feedback loops. Some will favor democracy. Some will favor autocracy. And they might all be changing at once. Some might be changing in concert as the result of carefully coordinated campaigns. Other loops might increase in strength in response. People don't always stand by when their voting rights are threatened. Sometimes they organize themselves and fight for more honest elections. That too changes the strength of key feedback loops within the system.

Activating dormant reinforcing feedback loops can create surprises. The most extreme changes to the relative strength of feedback loops happen when once-dormant loops are triggered into activity.

A snowball perched at the top of a hill isn't part of a feedback loop until it receives enough of a kick to overcome the friction holding it in place. Then it grows as it rolls downhill, picking up more and more snow. A citizens' protest might be launched by a whistleblower who witnessed an abuse of power and wrote an op-ed. That's an event. In response, a few people might gather at the city center; then more arrive, attracting still more. Even if the whistleblower never writes another article, the movement will continue to grow as members attract new members. Neither the snowball kicker nor the whistleblower matters once the reinforcing feedback loop starts feeding on itself. The citizens who might be mobilized don't resemble a movement before the whistleblower acts. The whistleblower doesn't look like a movement either, but between them, there's the potential for a system surprise.

The term *tipping point* is sometimes used to describe these sorts of reinforcing feedback loops, which once set off can carry on without additional need for whatever activated them. Sometimes activating events are consciously planned by people, such as the launching of a new idea or seeding of a movement. Some, however, are unintentional, like the release of greenhouse gases that warm the planet. As the planet's temperature warms, polar ice begins to melt. Because the color white reflects solar energy while darker colors absorb it, less ice means more of the sun's energy is absorbed by darker open water, leading to more warming, causing even more melting. The initial ice melt happened because of a linear chain of causation, from human-generated greenhouse gas emissions to warming temperatures to melting ice. But after enough of a kickoff, the warming could start feeding on itself. At that point, the loop would have enough momentum to continue even if human emissions were halted.

Triggering latent balancing feedback loops can change system behavior. Balancing feedback loops can also be latent within systems. They can be set off when something within a system shifts and creates a gap. They can be set off when goals change or when new goals are adopted.

But even though the "intent" of the newly activated balancing feedback loops may be to hold some goal steady, actions toward that goal can have consequences that ripple out to alter other parts of the system. For instance, a government might react to a financial crisis with austerity measures. They would close a gap in the national balance sheet, but other consequences would ripple outward, impacting social services, well-being, and infrastructure.

Distant causes and consequences can lead to systems surprises. Not every influence that matters within a complex system is located nearby. Not every event that matters happened a short time ago. Is a small farmer in Zimbabwe able to provide a good living for her family? The answer depends on many factors, which in turn depend on others. Rainfall amounts influence her crop. Climate change could influence rainfall, and energy infrastructure choices made decades ago in North America and Europe influence climate change. Choices made a continent away matter. But if we don't know about those choices, then the system's behavior registers as a surprise. The seed varieties the farmer uses will influence her yield. Local traditions stretching back generations might influence what varieties she plants. So might the marketing plans dreamed up at the distant headquarters of multinational agribusiness corporations during the last quarter.

Systems are vast and interconnected. To operate within such complexity, we have to decide what is important to pay attention to. If we tried to pay attention to everything, we'd spend all our time collecting information and making choices. To function, we have to explicitly define our focus and what our system's definition should be for our purposes, even though our decisions will have ramifications beyond.

Let's say I wanted to define the boundary of myself as a system. I could decide to consider my skin that boundary, imagining everything inside that mental distinction to be "me" and everything outside of it to be "the environment." But lately, medical science has been learning more and more about environmental pollutants, as well as about the boundary between me and "not me," revealing that it is fuzzier than many think. Endocrine disruptors, pesticides, and air pollution are all substances that seem to be outside of us yet find their way into us, with consequences for our health. Illnesses caused by environmental pollutants are systems surprises, which happen when a real system is more expansive than our mental map of it.

The boundaries we draw on both printed and digital maps are also mental maps. We create lines and say they define nations. But some aspects of the Earth, like the atmosphere, are fluid. The atmosphere doesn't abide by national boundaries. Forests in one region can suffer from air pollution from distant power plants. If we deal with governance systems, power plants, and forests separately in our thinking even though they are connected in the real system, the system known as the forest isn't likely to behave as expected. In the 1980s in the northeastern United States, we called this systems surprise "the acid rain" crisis.

Delays create surprising behavior if we don't account for them. In an effective balancing loop, harm will register as a gap between the goal (no harm) and the reality (harm), which subsequently will initiate action to close the gap. The behavior of the system will depend on the length of the delay between sensing the gap and acting to close it.

When you touch a hot stove, for example, the amount of time it takes for your brain to register the heat and send nerve impulses to move your hand matters. A quick response—a fast-acting balancing loop—will save you from injury; a delayed response could leave you with a serious burn.

Balancing feedback can get more complicated. What if its corrective action requires slowing down a reinforcing feedback loop? The gap is sensed. The corrective response is initiated. Both take a certain amount of time. Meanwhile, the reinforcing feedback loop is still spinning, amplifying the initial problem. The problem grows while the action to correct it is only beginning to ramp up. By the time the response "closes the loop," the system may have reached an entirely different state, which may require a different response, which also may be delayed, during which time the state of the system is continuing to change. . . .

Sometimes actors within systems try to increase the delays as a strategy to influence them. This is familiar to anyone who has worked to push governments to act on climate change. When emerging scientific consensus on the causes and consequences of climate change threatened the power of the fossil fuel industry (in essence, triggering a gap between the desired and actual security of the industry), oil and gas companies launched a highly effective climate denial campaign.[1] Sowing doubt and confusion, climate denial delayed governmental climate action, which itself was a balancing loop aimed at protecting the global climate. Increasing the delay in a goal-seeking balancing feedback loop can prevent that loop from ever accomplishing its goal.

Delays can also produce surprises when it comes to changes you are *trying* to create. Say your multisolving project aims to institute a new lower target for pollution in a lake; the stock of pollutants will eventually adjust to the new goal, but remember that while goals can change quickly, stocks don't change instantaneously. Even if the rate at which new pollution is added drops off, it could take time for the existing pollution to break down or wash away.

Maybe you build agreement in your community—including citizens, donors, and activists—that all families in the community should have easy access to fresh food. Fantastic! This new goal creates a gap, and efforts begin to close the gap. But it could take years to plant new gardens, start farmers markets, and build new food pantries. Your project is working, but you'll need patience. Don't let the delays in systems surprise you into giving up prematurely!

Systems responding to distorted information behave in surprising ways. Imagine trying to keep a room comfortable with a broken thermostat. Even though the room is overheated, the thermometer registers a cool temperature, signaling the furnace to deliver even more heat to the overheated room. When inaccurate information flows from one part of the system to another, the system's behavior will be distorted too. By sending falsified data to a regulatory agency, a manufacturer distorts an information flow. That flow is part of a balancing loop. The goal of the loop—to keep public safety at a certain level—can't be accomplished with inaccurate information.

Because distortion of information can have such dire consequences, some forms of it are illegal: corruption, bribery, perjury. But even milder forms of distortion can change the behavior of systems. Say a team runs a workshop, testing out a new design. It is a giant flop, but they report to colleagues that it went great. Maybe they are afraid of poor performance reviews, or maybe they are embarrassed that the new design didn't work out. What might happen next? Maybe others will adopt the new design, expecting good results, but end up flopping too. They'll waste time and disappoint clients.

Systems respond to whatever information is delivered by their interconnections. When the information is sporadic, erratic, or false, the behavior of systems will be distorted too. Promoting more accurate information can itself be a way to multisolve. Whistleblower campaigns, journalistic fact-finding, citizen science, and storytelling that reveals an injustice, each of these improves information streams. Each can help people recognize the actual state of their systems and register gaps between those states and the desired states. Good, accurate, timely information can be the seed of powerful balancing loops.

Surprises can come from a system's past. Two systems might contain the same elements, connected in the same ways. But if the two systems began operating from different starting points, they could end up behaving quite differently. Those starting points are called initial conditions. They include the relative fullness (or emptiness) of each stock and the initial rates of flows.

We understand this phenomenon about familiar systems. Let's say two businesses both exist under the same framework of rules and incentives. Both produce the same product, and both face the same competition. But if one business started off in debt and the other had a generous cushion of seed money, they would be unlikely to develop in the same way. The business with the seed money might be able to take more risks. It might be able to outlast a recession. Unless you knew about the difference in initial conditions, you might not understand why the businesses are so different today.

When you encounter a system at any moment in time, you are also meeting and engaging with its past. James Baldwin put it this way: "History is not the past; it is the present. We carry our history with us. We are our history."[2]

Even if two systems start with similar initial conditions, events over time, including chance events, leave their mark as well. This is called path dependence, and it too is familiar in our everyday experience. Two communities might have similar populations, in similar geographies, with similar economies. However, if an intense hurricane destroyed much of one community but bypassed the other fifty years ago, you might see traces of that chance event in the present. The hard-hit community might bear a legacy of trauma. Children who weren't even born at the time of the storm might be impacted by the struggles of their parents and grandparents. The building stock might be newer. There might be more poverty. Or a more cohesive community might have arisen from the shared effort of reconstruction. Without knowing the full path of a system, you may find yourself surprised at how it moves forward.

Partnering with people who hold the historical memories of the system (including histories of marginalized groups that may have been excluded from dominant narratives) in your multisolving efforts is one important way to hedge against being surprised by how the system's past still affects it.

When thresholds are crossed, a system's behavior can change. Some systems have the potential for sudden shifts in behavior. Thresholds contribute to this potential.

The amount of influence a change in one part of a system has on another part isn't always constant. A temperature increase from 25°C to 25.1°C may have a mild impact on a species' viability. But when the temperature increases from 26°C to 26.1°C there could be a drastic die-off. That's a threshold effect. Below the threshold, a small change in temperature leads to a small change in mortality. Above the threshold, a small change in temperature leads to a large change in mortality.

Not knowing where thresholds lie makes systems feel unpredictable, especially if all of your experience of the system comes from a time before it crossed the threshold. Because of threshold effects, understanding how a system works under one range of conditions doesn't guarantee you'll understand it under other conditions. As ecosystems are exposed to higher temperature extremes as a result of climate change, thresholds can be revealed as ecological impacts. As infrastructure is faced with higher temperatures or stronger storms, thresholds can be revealed as breakdowns or catastrophes.

Thresholds in one part of the system cascade outward to influence other parts. A breakdown in a supply chain for one product can result in delays in other industries. The threshold was in one system, but other systems are sensitive to it. Thresholds are unpredictable on their own. The ripples they can create are even more so.

Thresholds don't always imply danger though. Adding a hundred new supporters to a campaign can have a much bigger impact if it carries the campaign across a threshold for accessing campaign finance or debate participation, for instance. Or adding a few more acres to a conserved parcel might make the protected land large enough to serve as an effective refuge for an endangered species.

Because of emergence, changes in connections change how a system behaves. The interconnections of systems give rise to their behavior. If the parts of a system are connected differently, the system's behavior will change.

Here's an example that might be familiar to you. Picture two grocery stores, each with a fish counter. Each sells the same type of fish. Each posts the price per pound of the fish. But one grocery store puts little colored tags near the price, certifying that the fish was harvested by a sustainable fishery. They tell the shopper that the fish population is healthy and the fishing practices are sound. Red tags mean the opposite. The fish population is in decline or the fishing practices are unsustainable. Yellow tags signify something in between. With the card system, shoppers know more. Now any shopper who wants to can contribute to sustainable fisheries.

The creators of the card system have established a new connection in the system. Now shoppers are connected to distant fisheries with more and clearer information. And systems with new connections can produce new behavior. That's because a system's behavior emerges from its connections. Novel connections can deliver novel behavior because of the system property known as emergence.

One of the most powerful ways to encourage a system into new behaviors is to change the connections within it. This is a critical part of multisolving. New relationships between people are an example of such connections. Say a multistakeholder partnership connects organizations across different areas of expertise. Information and knowledge flow. Resources are shared. Maybe one organization has a star grant-writer who coaches the other organizations. When one organization has a key hearing at city hall, all of the other organizations can show up in support. When all the groups act together, the system will respond differently than if each acted alone. That's emergence. Because of new connections in the system, new behaviors emerge.

Emergence is another source of unpredictability in systems. You can understand the parts of a system well, but that may not prepare you for what happens when those parts are connected in a particular way. When I think about emergence, I think about water. Water molecules, of course, are made out of two hydrogen atoms and one oxygen atom. Hydrogen and oxygen are gases at room temperature. Water, a liquid, emerges when those atoms are connected by chemical bonds. You can't understand water by studying hydrogen and oxygen separately. You can't understand a forest by studying maple trees, pine trees, squirrels, and earthworms individually either. Some properties of forests emerge from the interactions of the parts.

The "magic" that sometimes happens in teams is another example of emergence. You can know that Ron is good at spreadsheets and has a funny sense of humor, Susan tends to think before she speaks, and Maria jumps into new projects with full enthusiasm. But until you put those three individuals together into a team, you can only guess about how the team will perform. Will Maria's eagerness balance out Susan's thoughtfulness, or will their styles create conflict and tension? Will Ron's sense of humor smooth out the mix? "Team-ness" does not exist in the individuals. It exists in the connections between them. That is a hallmark of an emergent property.

Networks are emergent systems: people connected in a social movement, neurons in your brain, businesses in an economy. The memory held in your neural networks doesn't exist in one particular place; it is an emergent property of how the neurons are connected. The winning coalition strategy doesn't belong to one individual; it emerges out of a wild brainstorming session between them. Keep in mind, however, that systems change when connections are broken too. When one species is lost, a whole ecosystem might shift into a different mode. When one person leaves a team or when two people on a team have a falling-out, the team changes, maybe subtly, maybe drastically.

Emergence is a powerful force we can nurture for multisolving, leading to innovation, creativity, and new possibilities. It also should keep us humble. Nobody can really be sure what will happen next when the connections in a system reorganize.

Systems are sensitive to the influences of both smaller and larger scales. Systems both are composed of subsystems and exist within larger systems. Cells are systems within organs, which are systems within organisms. Neighborhoods are systems within cities, which are systems within states. Sometimes systems are not sensitive to changes at other levels. But sometimes changes at one level of a nested system can influence dynamics at other levels.

A new state law could make it easier (or harder) for cities to expand their public transportation. In that case change at a higher level of system organization ripples downward to influence systems at a lower level. The opposite is possible too, of course—change at a lower level of nested systems can ripple upward. A new mayor in one city might gain broader prominence because of city council's bold innovation. That mayor might rise to become governor and implement some of the same innovations she experimented with as mayor statewide.

Nested systems are a daily feature of our lives. A missing enzyme in my cells might manifest as liver disease, which would negatively influence my body's energy and vitality. That could ripple upward to influence my family system, as other family members help care for me or take on tasks I can't do and as I take on other roles and contributions as I am able.

The influences rippling from other levels of nested systems can be sources of unpredictability. We may not control or even influence these other levels. We may barely be aware of them. But they can have tremendous impacts on the systems we are participating in or trying to manage.

Deeper Influences

Our choices and decisions, belief systems, and cultures all influence the structures of the systems we create. Teasing apart how these influences shape systems, however, is difficult. Different systems theorists use different terms for these influences, suggest different groupings, and have differing ideas about how they interact.

I have my own take. The view that makes the most sense to me is shown in the bottom two layers of figure 6-1. It synthesizes the work of several different thinkers, including the "iceberg" metaphor used in many systems

thinking classes; Donella Meadows's work on vision and my own expe-
riences with it; the work of Nathaniel Smith at Partnership for Southern
Equity on values-based organizing; and the work of Mallary Tytel and Royce
Holladay in their book *Simple Rules: A Radical Inquiry into Self* and Glenda
Eoyang and colleagues at the Human System Dynamics Institute.[3]

Systemic structure is in constant flux. Some physical parts are wearing
out and being replaced. New physical elements are being added on. We
recognize this in familiar systems. Your body is constantly adding new cells
and shedding old ones. A city is constantly adding new buildings and tearing
down old ones, writing new ordinances and discarding old ones. At any one
time the structure of the system—and thus its behavior—depends on all of
the past alterations of systemic structures.

Of course, the system is reshaping itself all the time as well. Feedback loops
are turning. Processes, despite delays, are churning along. And events are
impinging: natural disasters, decisions made by governments or corporations, new scientific discoveries, market forces, fads, and more.

The people within systems are constantly changing them too. Whether
it's going along with the momentum of a system or organizing to radically
redesign it, we participate in a continuous shaping and reshaping of systems.
We write laws or repeal them, or at least our elected representatives do. We
participate in fads and trends and shift market dynamics in the process. We
change our thinking when scientists make new discoveries or storytellers
hold up a mirror to our world. All of this changes the structure of systems.

Moment to moment, the people shaping and reshaping systems are
influenced by three things that have deep and persistent impacts. These
influences are present at the ballot box, in the updating of the university
curriculum, in the grocery store aisle, and in the corporate boardroom. We
don't always have the power or the opportunity to shape a system, but when
we do, these three influences guide our actions and our choices. The first
influence is vision: What are our deepest hopes for the future? The second is
values: Who and what matter? Simple rules—organizing principles that are
applied (often unconsciously) across different parts of a system—represent
the third influence.

Vision. When a system renews itself, when choices are to be made, vision—
our picture of a desired future—can have a strong and deep influence on the
choices we make. Those choices change the structure of the system, perhaps
in small ways, perhaps in profound ones. And since a system's structure gives
rise to its behavior, visions of the future end up influencing the behavior of

systems in the present. A vision of our world without hunger, if deeply and widely shared, would change economics, agriculture, land use, education, and food distribution.

Values. If the people involved in the constant shaping and reshaping of systems value the rights of nature, they will make different choices about tax rates and conservation hectares than they would otherwise. Do they think some groups of people are more deserving of safety or health or wealth than others? Those values too will influence choices and decisions that shape systemic structure. Which stocks' flows and feedbacks will change? How much? How quickly? As values shift, many different aspects of a system shift too. The value "who matters" widened at least slightly in the United States because of the Civil Rights Movement. Voting laws changed, which in turn impacted who won elections. Civil rights legislation changed the workforce and education and housing. Of course, we still have a long way to go, as evidenced by remaining racial inequities.

Simple rules. In addition to vision and values, simple rules are also important at moments of choice and change. If employees accept "be experimental" as a core organizing precept, the system will end up with a different structure than if they adhere to "do whatever the boss says." If they adhere to the rule "use only sustainable materials," after some years the system will look very different than if they had applied "use only the least expensive materials."

Even when only slightly adjusted, vision, values, and simple rules can set off change in many parts of systems at the same time. This makes them targets of intervention within systems, including for multisolving.

There is an even deeper influence on systemic structure, however, from which vision, values, and simple rules all emerge: worldview. Sometimes referred to as paradigm or mindset, a worldview is a shared theory of how the world works. Is it a series of mechanisms? Is it a living and interconnected web? Is it a collection of independent objects? What are our responsibilities to one another? Answers to questions like these make up a society's worldview.

Worldviews aren't constant, though they can be slow to change. Some worldviews that were once common, at least in some cultures or parts of the world, are no longer adhered to today. For instance, few Europeans today believe that the sun orbits the Earth or that kings are anointed by God. Because vision, values, and simple rules each derive from worldviews, changes to worldviews—like the adoption of heliocentrism—can cause dramatic changes in systems. Vision, values, and simple rules already touch many parts of a system at the same time. Shifts in worldview influence all three, and that can lead to surprising and far-reaching changes in systems.

Vision, values, simple rules, and worldviews are often implicit, unlike the physical stocks of systems—the cars, planes, grain combines, forests, bicycles, and shopping malls. While many of the connections within systems—the property tax rate, the cost of an acre of land, the quota per boat in the cod fishery—are complicated and technically intangible, at least they are concrete. Vision, values, simple rules, and worldviews are not concrete. It is easy to leave them unexplored and unarticulated. Sometimes these deep influences within systems are so widely accepted that we hardly notice them. Sometimes conversations about them are so charged that we shy away lest we ruffle feathers at the dinner table or offend powerful constituencies in the community. But our lack of awareness of these deep influences provides yet another reason why systems surprise us. Some of their deepest drivers are invisible.

Why are all the feedback loops operating *this* way? Why are balancing loops aimed at *these* goals? Why is this stock (of wealth or water or power) so high while this one is so low? Why is *this* link in a feedback loop so weak? Why is *that* delay allowed to persist? Often the answer to questions like these will be found in the vision, values, simple rules, and worldviews of the people with the most influence in the system. And of course, even the question of who has the most influence is shaped by vision, values, simple rules, and worldviews!

No wonder systems surprise us. No wonder they produce harmful behaviors or resist our change efforts. It's not just all the intuition-fooling complexity. Systems are also responding to deep and sometimes unexamined influences. Learning to recognize these influences and examine and change them is another part of the art of working with systems.

Because vision, values, simple rules, and worldviews affect so many parts of systems, they are of particular interest to multisolvers. One value, if shifted, could restructure many parts of a system, addressing many problems at the same time. Think how much would change if the value "clean water is sacred" came to be commonplace in industrial societies. Agriculture might conserve water used for irrigation and stop polluting waterways with nutrient runoff and pesticides. Households would also use water more sparingly. Cities might capture runoff from rooftops and roadways, purifying it through rain gardens and other living systems. Standards of allowable pollutants would change. Coal, timber, oil, and other resource extraction practices would be affected or even eliminated. And so on. That one value shift would be transformative.

Table 6-1 offers a few more shifts in vision, values, simple rules, and worldviews that would profoundly alter the systems we live within. For each of them ask yourself if you know of small systems wherein they are already strongly held and what would happen if they were adopted on a wider scale.

These are just a few examples that occurred to one person. Challenge yourself to imagine as many alternatives to dominant values/visions/rules/worldviews as you can and what kinds of ripples of multisolving they would unleash. As you read my list or create your own, be curious about your own reactions. Do these alternatives seem impossible, impractical, naive? Is there a voice in your head saying "that will never happen?"

These deeper layers of systems can feel immutable. Yet although less tangible, less discussed, and often serving the interests of powerful vested interests, these parts of systems—unless they defy laws of physics or nature—are still "fair game" for change efforts. Imagining alternatives and experimenting with them on smaller scales can be a way to steer systems toward new possibilities.

Table 6-1.

VALUES	VISIONS	SIMPLE RULES	WORLDVIEWS
Every child matters.	A multiracial democracy within my lifetime.	Keep money out of politics.	The purpose of life is to give back.
Women and men are equals.	No cars in the city center within five years.	Share wealth.	The best way to be safe is to protect the most vulnerable.
Nature has rights.	Permanent protection from unstainable extraction for the world's oceans. Worker-owned clean energy infrastructure.	Exercise precaution.	The laws of economics should be derived from the laws of the Earth.

Sometimes the majority of people participating in a system adhere to the same visions, values, simple rules, and worldviews. Whether implicit or explicit, there's a high degree of agreement about how the world works. The image of a desired future, what matters, and how things ought to be done emerge coherently from that worldview. In other systems multiple visions, values, and worldview can coexist at the same time. Even while colonial empires swept the world, Indigenous peoples survived and continue to survive and embody very different worldviews than the one that justified conquest and empire. Adherents to a particular worldview may have more access to system-shaping power, but that doesn't mean opposing worldviews vanish.

During times of change, when new understandings are emerging, prevailing worldviews can be in flux as well. Confidence in one worldview can ebb while confidence in another rises. What does all of this mean for the structure of systems? During the constant modification and upkeep of system structures, the people making the choices that shape systems may be responding to different sets of visions, values, simple rules, and worldviews. One part of the system might be modified by someone holding a particular vision or value. Another part may be changing under the hand of a group ascribing to a different vision. A structural element put in place under one vision might be revised by the next influence holder, under sway of a different vision.

That brings us to the issue of the power to influence systems. Who has the power to enact the most substantial changes in the systemic structure? Who has access to determine the most critical rules, laws, and incentives? Who is influencing economic policy or shaping the stories that are broadcast nationally? Who is designing educational standards or setting energy policy? In chapter 10 we will look more deeply into issues of power and equity. It matters who shapes systemic structures and what worldviews, visions, values, and simple rules they espouse.

Embracing Complexity

We are embedded in complex systems. We don't have any choice about that. Complexity and interconnection are a feature of our world. What might be possible if more of us, and more of our leaders and organizations, embraced that complexity?

What does it mean to embrace complexity? What would change in our lives and work and leadership if we did? I think there are at least three ways.

The first is system awareness. You see the world differently, as complex and interconnected. You ask questions about a system's past, its momentum, and its interconnections. You look for the cause of undesired system behaviors in the structure of the system itself. You become more conscious about where the boundaries of systems are being drawn in your mind and in the minds of others. You look for unspoken vision, values, simple rules, and worldviews.

I hope that the material in this book so far has helped you cultivate some of that awareness. I hope that, even if that's all you take from this book, you'll never see the world in quite the same way again. And I hope that this perspective will help you see more and understand more about tumultuous times and converging crises, wherever you may encounter them.

The second way to embrace complexity is to get better at navigating systems. This goes beyond seeing the world in systems terms. To navigate systems, you act within them, trying to achieve better results, and trying to avoid unanticipated side effects. You surf the waves of tumultuous times. You ask questions and change strategies based on the answers. How might the system push back? Where is momentum carrying us? You think about key stocks and make decisions based on what you discover.

If more of us become adept at navigating the complexity of systems, we will have better outcomes in this time of overlapping crises. We'll anticipate and avoid at least some trouble. We'll save lives and preserve options for the future, from protected ecosystems to resilient neighborhoods.

But from where I sit, being aware of systems and being skilled at navigating them is not enough. With problems growing in magnitude and gaining in momentum, we need to go beyond surfing the waves to transforming the systems' structures that are creating the waves in the first place.

That brings us to the third way to embrace complexity: steering systems, which requires creating fundamental and lasting change in systemic structures so as to create new, healthier system behaviors. To steer systems requires more than just awareness and navigating, it requires changing the elements of systems, how they are connected, or the visions, values, simple rules, and worldviews that influence the choices of those who are shaping systems.

Steering systems isn't easy. We've already seen many of the ways that systems' own momentum and the influence of powerful, vested interests cause systems to resist change, but multisolving can help ease the challenge of steering systems. By clustering problems into bigger wholes, we can draw on more people, more resources, and more creativity to enact change.

Multisolving can help us know where to intervene and how to steer as well. It can help identify strategic points of intervention. If some part of a system is linked to multiple problems, then it's probably a fundamental and significant part. Determining that a system change could benefit many goals helps you be sure that it's a strategic point of leverage.

In the chapters to follow we'll look more deeply into embracing complexity. In the next chapter we'll look at navigating complexity and how doing so is a part of multisolving. Then, in chapter 8, we'll explore steering systems using multisolving. What attitudes and approaches are helpful? What do networks have to do with system change? And can multisolving be at once a way to address multiple problems and a way to seed a different set of visions, values, simple rules, and worldviews? Could multisolving address crises while also giving us a chance to practice a different way of being in the world than the way that generated so many crises in the first place?

? Questions for Reflection

- Think of a time you took an action and got a substantially different result from what you expected. Write down what you did, what you expected, and what actually happened. What did that surprise teach you about that system?
- Think of a system you are involved with right now whose behavior catches you and others by surprise, is counterproductive, or is even dangerous. Look back at the section "Whole System Behavior." Do any of the phenomena described in that section seem to be at play? Make a list. If you have an opportunity, try explaining your list to someone else who knows a lot about that system.
- Write about a time (or better yet, several times) you experienced emergence, where something novel happened because a new connection was made within a system. Possible topics: a new idea emerged from a conversation or a brainstorm; a new partnership between individuals or organizations accomplished something; adding (or subtracting) an element changed a result in the kitchen, or a garden, or an aquarium, or an ecosystem. After you reflect on the specific examples, write about what emergence feels like. How do you know it when you see it?

- Where do you feel personally aligned with the visions, values, simple rules, and worldviews of systems you participate in? Where do you feel a lack of alignment?

She Stands

She stands,
poised,
on the balls of her feet
while around her everything twirls.
A still point
made of movement.

CHAPTER SEVEN

Rising to the Challenge of Complex Systems

Given that the behavior of complex systems is so full of surprises and difficult to control, is it all hopeless? Should we give up in the face of interconnection, delays, distortions, shifting feedback loops, thresholds, and all the rest?

A lot of people do give up, of course. Some decide that systems are too complicated to understand, let alone influence. They adopt a kind of fatalism: "Who can know what will happen next?" "It's all rigged by the powerful anyway and nothing much ever changes." "Better to simply live your life and take your chances."

Others try to get along by walling off a lot of the complexity. Consciously or unconsciously, they choose narrow system boundaries to keep things "manageable." They focus on this year's finances but not the implications for their grandchildren's economic prospects. They track the cost of raw materials to the bottom line of the company but not to the ecosystems that produced them. In fact, many laws and financial incentives are crafted to facilitate such a narrow focus.

Looked at one way, a lot of successful managing goes on within these self-defined boundaries. Products are made. Discoveries happen. Technologies advance. For many people, quality of life improves. Yet outside of these carefully drawn boundaries, complexity continues to operate. Wastes seep outside of our well-managed boundaries and pollute the groundwater and the atmosphere. Living systems show the scars of extraction. Groups of people whom the rules of the economy count as beyond the part of the system that "matters" bear the brunt of the damage. Global cycles of water, carbon, and nutrients tilt out of balance.

We see the impacts in some of our largest shared systems, but the price of ignoring complexity shows up in our communities and organizations too. Maybe sprawl has you stuck in traffic. What has decades of urban planning considered "outside of the system" to produce the specific configuration of housing, highways, and traffic congestion in your city? Which communities mattered and which ones didn't? Which communities does the metro rail serve and which does it not, and how did wealth, race, and power help create the system? No one set out to trap you in traffic, but that's the result of complexity ignored by decision-makers.

Maybe tensions are high at your company between the marketing department and the product designers. What feedback loops between the two are producing perverse results that nobody wants? What incentives and flows of information (or lack of flows) are creating the dynamics?

Not engaging with complexity doesn't seem to be working, so what are the alternatives? Are there ways to approach complex systems that decrease the odds of them surprising us? If we can't fully rid systems of their potential for surprise, can we be better prepared for a world of surprises? Can we find ways to live and work within systems that are less likely to cause harm and more likely to heal it?

After years of teaching systems thinking, based on the wisdom of many of my systems teachers, my answer to each of these questions is an emphatic yes. The good news about the poor grasp of complexity that we see in some of our most important systems is that there's huge room for improvement!

A Systems Stance

Those who study and teach about systems have, for decades now, been offering suggestions for how to conduct ourselves in a world of systems. One of my favorites is Donella Meadows's list in her essay "Dancing with Systems."[1] And Peter Senge offers in his book *The Fifth Discipline* a different list of traits that are particularly useful in what he calls a "learning organization."[2] I offer my own list in this chapter, influenced by both Meadows and Senge and by my own experience.

Taken together I think of the following list of seven suggestions as a systems stance. It offers a way to participate in a world of complexity. And because multisolving always happens within the context of complexity, these suggestions—less hard and fast rules than starting points for experimentation—also represent ways to increase the odds of successful multisolving.

Try them out. Decide for yourself if they help you achieve better results, help your systems work better, and allow you to navigate tumult and get your bearings in the midst of disruption.

Visualize systems. There are many ways to look at the world. Binocular vision helps you see depth. Magnified vision helps you see detail. The first part of a systems stance could also be called "systems vision." You can visualize systems if you try, and improve if you practice. In chapter 2, for example, we imagined looking down from a mountaintop at a watery landscape of stocks connected by flows. That's a form of systems vision.

Here are some questions you can ask to sharpen your systems vision when considering systems that matter for you and your work:

- Where have we drawn the boundaries of this system? What would change if we drew them wider (or narrower)?
- What about our boundaries in time? What is the time horizon of our decisions? What would change if we considered longer (or shorter) time frames?
- What are the key stocks?
- What is flowing? What needs to flow?
- How is information flowing? From what to what? Cleanly or with distortion? In a timely fashion or with delays?
- How do the problems you are bringing together for multisolving relate to one another?

When you encounter an unfamiliar system, systems vision allows you to use systems you know well as analogues. As we ourselves are complex systems living in a world of systems, we all have a library of systems we can draw on in novel situations.

Imagine that you've been tasked with recruiting hundreds of new members for your organization when you only have a few dozen right now. You know reinforcing feedback drives change that starts slowly before increasing precipitously. What does your campaign have in common with a snowball rolling downhill? How might you get such a process moving in your situation?

As a young graduate student, I volunteered with a national grassroots movement called Beyond War that was concerned with peace and nuclear disarmament. Beyond War used a train-the-trainer model, conducted almost entirely in people's living rooms. I was invited to an introductory presentation in the living room of a couple who had relocated their young family to my state to try to build a new "chapter" of Beyond War. Talk about trying to seed a reinforcing feedback loop!

I remember being impressed at that presentation. Wanting to get more involved, I went through a series of discussion sessions. Then I was invited to a facilitation training, where I learned how to host the same discussions I had just participated in. For support I had flyers and scripts to help me invite friends and acquaintances to the gatherings I hosted. I traveled out of state to national meetings to meet other volunteers and came back energized. Looking back, I can see how the leaders of that movement had given careful thought to the dynamics of reinforcing feedback, putting supports in place for when leadership would come from new members, then the new members those new members touched.

Cultivating systems vision means that you take the time to identify important stocks and flows and feedbacks. You ask where the system is coming from and where its momentum might be carrying it. You get curious about initial conditions and path dependence. Cultivating a systems vision isn't something you do only once at the beginning of any project or initiative. It also requires you to pay attention to how systems evolve. Are feedback loops shifting dominance? Does that matter? Are new connections being made? Do you see any signs of new behaviors in a system?

Develop a systems ethic. Recognizing that systems are complex, interconnected, and interdependent has implications. Taken together, these implications point toward ethical principles that can be helpful guides in the midst of complexity and rapid change. When in doubt, check in with your sense of systems ethics.

The precautionary principle is part of a systems ethic.[3] Recognizing the unpredictability of complex systems, it urges great care when intervening— don't change things if you don't understand the system! It advocates careful and thorough testing of new innovations. Whether adding a new chemical compound to a food supply or a modified organism to an ecosystem, the precautionary principle calls for moving forward with proof of safety in hand before action is taken. This part of a systems ethic is an important hedge against the potential for unintended consequences.

The golden rule, taught in preschools and in wisdom traditions, also fits well in a systems ethic. "Do unto others as you would have them do unto you." Complex systems have long chains of circular causation. Through these linkages, what you do unto others may one day feed back to impact you. Imagine if everyone, not wanting the water they drink to be polluted, refrained from polluting all water. Imagine if everyone, not wanting the impacts of wars in their homelands, sought peaceful means to resolving conflict. The golden rule is a good reminder to expand the boundaries of "the system." It is a reminder to consider, before acting, the "other" whom you might have conveniently imagined as outside of "your" system.

The golden rule doesn't always have to be an exercise in imagination. A system ethic also includes the concept of "asking for consent." This is the basic ethics of inquiring, in advance of an action one might take, for permission from another. When systems are large and complicated, we might not have enough information to "do unto others as we'd have them do unto us." When possible, we can ask. That's consent, an idea that is familiar in the realm of dating and sex but that can be applied more broadly. Does the developer of a new sports stadium seek consent from the neighborhood that will be faced with traffic and noise? Does one country ask consent of all the countries downriver before diverting water for irrigation? The idea of consent could be taken even further, given recent work on the rights of nature and the rights of future generations. Do people yet unborn consent to the changes current generations are making to the climate? Does the ocean consent to having plastic dumped into it?

We are so used to a world that doesn't operate from a systems ethic. For that reason, these proposals may sound unrealistic or idealistic. If you find yourself thinking, "Sounds nice but it will never happen," I suggest a thought experiment. Think about something that concerns you or impacts you. Would that crisis be abated, at least a little bit, had a systems ethic been in place? Maybe it's not a systems ethic that's unreasonable. Maybe it's the belief that we can keep on managing within complex systems without one.

Prioritize learning. It's hard to be sure how systems will respond to our interventions due to their unpredictability. We may have pretty good ideas about short-term impacts, but the further out one gets, the more uncertainty there is. We may not know if there are thresholds built into the system. There may be side effects we don't intend. It's challenging, if not impossible, to prepare for every possible outcome.

Still, often we must act. We may not have a choice but to move forward in complexity and uncertainty. Inaction also influences systems, of course, producing ripples, pushing systems over thresholds, and inducing side effects. A learning stance offers a way to meet this challenge. You start out being as clear as you can about your expectations. What's your desired end result of your intervention? What's the state of the system you want to achieve? With these goals in mind, you can then act, and later investigate what actually happened, to the best of your ability.

Taking a learning stance means doing this investigating and reflecting again and again, each time you act. You assess progress (or lack thereof) toward your goal, for both big and small efforts. How did that massive million-dollar infrastructure project turn out? After a challenging discussion with a colleague, you ask them, "How was that conversation for you?"

My friends with military training call this process the "after-action review." It will help you see your influence on systems. It will show you parts of the system that you may not have understood. The answers, in turn, will help make your next action more effective.

Say you just completed a multisolving intervention, but you honestly can't see much change in the system. What to do? Investigate. Did change happen initially and then dissipate? Maybe there's a balancing loop pushing against the change you made. You can look into that and design awareness of it into your strategy going forward.

Perhaps you're in the middle of a project but aren't seeing change yet. Could there be a longer delay built into the system? You might search for earlier indicators. What would be the first thing to change—can you investigate that? Or is change simply building more slowly than you expected? Are you able to keep watching and keep learning?

If you do see changes, you won't always know that your actions were the sole cause. In fact, in a complex system, it's most likely that your actions were *not* the only influence on the outcome. Remember, in systems there are many influences all operating at the same time. A learning stance includes paying attention to context.

My friend Tina Anderson Smith works with leaders in health and sustainability projects to help them learn and document their learnings. Her goal in complexity evaluation is to move from a description of whether something happened to an explanation of how something happened in particular ways and conditions. She has a good way of describing this process of embedding your learning in context. "You don't want to know whether X 'worked,'" she says. "You are looking to sharpen your understanding so that you can say, when A, B, C conditions are in place and we do X, Y is likely to happen."

This type of contextualized understanding will help you take a "success" in one area and replicate it in another. To do this sort of learning well, you may need to focus your attention on several levels at once. You need to look with a sharp focus for the changes you are trying to create, but you'll want to step back with a wider view to scan for other ripples that you might not have anticipated. Sometimes a surprising ripple of impact will change your understanding or even pivot your effort in a whole new direction.

Tina and I first worked together on a project in Milwaukee, Wisconsin, where we used a computer simulation to help leaders test strategies to reduce urban flooding. The simulation tracked changes to different variables—jobs, costs of labor and materials, flooding, water quality—under different strategies. Our goal was to help leaders evaluate a full picture of costs and benefits. Tina's evaluation showed we were able to help with that. But as our work progressed, Tina noticed something else. The idea of balancing co-benefits began to become a theme in the discussions. It started when stakeholders came together to talk about the simulation. Then it began to show up in other venues, like municipal meetings. The simulation exercise participants Tina interviewed reported that although the idea of co-benefits wasn't something they had thought of before, it was now influencing their work, leading them to form new partnerships and look for win-win-win solutions.

Tina and I learned that we weren't only informing decisions about flood prevention. We were influencing how people thought about synergies and collaboration. This observation changed how we conducted our project work in the next city where we worked. We spent much less time analyzing flood prevention scenarios. Instead, we invited conversations about partnership and synergy. We designed meetings to help new connections grow between stakeholders. Five years later, the project in the second city continues to grow and evolve, having launched partnerships, changed policies, and supported emerging leaders. It all started by looking for impacts of our work beyond the ones we expected.

Tumultuous times are pushing systems into new zones of operation. That's another reason learning is so important. We are operating in new ecological conditions. Economic conditions are shifting. Old cause-and-effect relations may not hold out anymore. Still, you have to act. You'll need to make choices and plan strategies. Experiment. Pay attention to what happens. Share your results so that others can learn. All of this will help both you and others be nimbler and adjust more easily to changing conditions.

It takes resources to learn. It takes time and attention. Sometimes it requires creating and filling new roles in projects and organizations. In steady times, where what worked before is likely to work again, learning might be a "cost" you can afford to skimp on. But not in times of rapid change. If you want to become ever more effective, invest in learning. If you need to navigate in unfamiliar terrain, invest in learning.

Simulate scenarios. Tools exist that can help us explore the potential consequences of our actions within complex systems. One of the most helpful is computer simulation, especially system dynamics simulation.

Computer simulations use mathematical equations to keep track of important stocks, flows, feedbacks, and causal connections in a system so you don't have to. They make it easy to test different strengths of connections for key relationships. They allow you to start the system off with different initial conditions and see how much that matters. They allow you to simulate different interventions and see how the system responds.

Computer simulations also run very quickly. This allows you to test in seconds or minutes scenarios that might take decades or centuries to play out in the real world. We don't always have time to conduct real-world experiments. Sometimes the consequences of bad choices would be devastating. Citizens and leaders can use simulations to build their understanding of complex systems more quickly and with little risk. Having learned in the simulated world, they can be more effective in the real one.

Climate Interactive creates climate change simulations. Thousands of people around the world, from secondary school students to climate negotiators, have used Climate Interactive's simulations to test scenarios. What if China reduced its emissions in 2030? What if there were no more loss of forests worldwide? What if a global price on carbon of $200 per ton were established in 2035? And so on.

Climate Interactive's modelers write the mathematical equations required to answer such questions. The equations track stocks of greenhouse gases and power plants and vehicles. They represent feedback loops in the economy and the Earth system. Others on the team create interfaces that allow users to set parameters and move policy levers to ask their own "what if" questions.

System dynamics experts elsewhere have built simulations of all sorts of systems, including production lines in factories, pandemics, chronic disease, predator/prey dynamics, and commodity systems. My own intuitive feel for complex systems has been improved by building and using computer simulations. It happens quite often that a simulation result doesn't match my mental model, giving me a chance to learn. What stocks are changing at what rate? Which feedback loops strongly influence the results? Which seem to matter not much at all? This sort of experimentation has taught me about specific systems but also about systems in general. It has taught me that my hunches are not always right. It has taught me that there's often more than one reason for why systems behave the way they do. It has taught me that there are windows of opportunity when systems can shift and points of no return beyond which they flip into new behavior modes.

Today there are excellent software packages and training courses that help you experience such simulations.

Giving decision-makers in government, corporations, and communities more access to computer simulation is a good investment for tumultuous times. As voters and stakeholders this is an important resource we can know about and advocate for.

Still, people won't always have the time or resources to draw on computer simulations. You still can engage in scenario thinking, though. You can ask "what if" questions. You can draw on what you know about the stocks, flows, and feedbacks of the system to answer "what if" questions. Even better, you can engage in scenario thinking as a part of a group in which each participant has direct experience with a different part of the system. Collectively, it's likely you already know a lot about how the system might respond to different scenarios.

Build your tolerance for uncertainty. Even if you're committed to learning and scenario thinking, there's still going to be a lot that you don't know when intervening in systems. Under these circumstances, it is important to build your tolerance for uncertainty. Can you act with limited information? Can you get a good night's sleep after making a decision whose outcomes you can't possibly know? It's not always easy, but a tolerance for uncertainty is something that you can practice and build over time.

In complex systems, the quest for certainty is actually futile, for many of the reasons we explored in the last chapter. Knowing this actually helps bolster my own tolerance for uncertainty. It helps me to remember that no matter how hard I try, I can't have certainty, so why waste precious time and energy chasing it? That energy would be much better spent testing scenarios, experimenting, and learning.

Sometimes the quest for certainty is more than just a misplaced use of time and energy. Sometimes, it can be limiting and dangerous. Think about the type of leader that many societies and organizations still today tend to elevate: someone projecting strength and certainty. Someone who brooks no questions, no hesitation. Often an older white man. Such leaders don't say, "I'm here to tell you times are uncertain, I have many questions, I'm committed to experimenting and learning." When a criterion for leadership is certainty, it's hard to be a learner. But if systems truly are full of surprises and difficult to control, leaders are no more certain than the rest of us. They, and all of us who depend on their decisions, need the freedom to experiment, learn, be uncertain, and change direction.

Being comfortable with uncertainty can be a collective pursuit. We can support each other's tolerance for uncertainty. If you're in a leadership role, you can reward figuring out more than knowing, experimenting more than predicting. You can create a culture where it's okay to be unsure of the right answer. You can admit that you aren't sure either. When your teammate says he just doesn't know but has an idea of how to test some possibilities, ask what help he needs. When your boss says she was wrong, recognize that as the sign of strength that it is.

One final important point about tolerance for uncertainty: it applies to your own thoughts, beliefs, and assumptions as much as it does to the rest of the world. Do you really know that "he's just out for profit" or "maple trees can never adapt to warming temperatures" or "they'll never listen?" Even when what you know feels like the voice of hard-won experience, remember that systems surprise us, the past doesn't always determine the future, and there's an awful lot we just don't know.

Play with timescale. In many systems, often without much thought, one timescale will dominate decision-making. Often, in mainstream systems, this timescale is very short. In corporate life it might be the quarterly earnings report or the annual profit and loss statement. In an organization it might be a hyperfocus on the next big gala event. With this timescale dominating, life may feel like a series of leaps from one event to the next. Longer-term goals might hover out there somewhere. But they never quite get their due of time and attention.

In tumultuous times, crises also tend to shorten time horizons. People are forced to prepare for, manage, and recover from disasters and destabilization. Short-term pressures loom large. But of course, in tumultuous times long-term dangers also threaten. They also need attention.

There is an alternative, though, and it is one that multisolving helps with. It is possible to cultivate a split attention. You can have one eye focused on the near term and one on the more distant horizon.

Within systems, our actions have short-term and long-term impacts all the time, whether we are conscious of this or not. The trick is to be aware of this fact and to be looking and thinking about more than one timescale at the same time.

If you look for them, most systems are full of opportunities where these two perspectives are synergistic. In fact, multisolving is, in part, the search for policies and investments that produce a needed result in the near term while also reducing risk or creating benefit in a more distant future.

You can wean a society off of fossil fuels to help prevent long-term climate change. But you can do it in a way that has benefits for the short term too.

Air pollution and the health issues it causes will fall as less coal, oil, and gas are burned. New jobs can be created in the near term.

Retrofitting buildings to make them more energy efficient is a policy with lots of near-term benefits, from jobs to lower energy bills. By reducing greenhouse gas emissions, it also creates long-term benefits.

Another element of playing with timescales is to keep one eye on the past. Remember the sensitivity of systems to initial conditions and the paths that systems have traveled? As you act within systems, can you look forward and backward simultaneously? Can you see the traces of past decisions, including inequities, in the present? Are amends or adjustments needed to make the system just and healthy? Is there a way to steer toward the future that includes making up for the violence and injustice of the past?

Remember that some of the system's history is embedded in physical stocks. The past may be seen, here in the present, in the placement of power plants, community wealth, and the geography of bus lines. Multisolving includes addressing the legacy of injustice in the past as part of the strategy for solving problems in the present or problems that loom in the future.

Play with spatial boundaries. Systems can be sprawling and vast. In the last chapter we discussed the temptation to superimpose narrow boundaries on systems. The system ends here, we say, at the city limits, with the company's profit and loss statement, or at the headwaters of the river. Sometimes this simplification works well. You can make agricultural plans based on how many cubic feet of water pass through the river each day. You may be able to ignore weather patterns over a distant ocean. But if winds, clouds, and rainfall over land shift as a result of changes to those distant patterns, your plans might suffer. They could be undone by forces you thought of as far outside the system.

Sometimes simplifying our mental maps of systems is a way to be efficient and move forward without having to worry that every distant detail needs to be accounted for in our plans. Sometimes we need to simplify, lest we be overwhelmed by complexity. However, there's always a risk when we do so.

One way to work with this risk is to grapple with boundaries. Be conscious of them. Be flexible with them. Visualize the systems you care about with your mental boundaries drawn in different places. See if anything changes about your strategies when you view the boundaries differently.

In your multisolving, one way to find solutions that link multiple problems is just this sort of mental examination of spatial boundaries. A toxic substance is a problem for people living near this manufacturing plant, but what about those living downstream? What happens when the substance is carried by wind currents? Maybe there's already a group protecting the invertebrates that are taking a hit downstream or another community being impacted by the prevailing winds. Expanding the system boundaries in your mental maps expands your mental landscape of potential partners and allies too.

You can find new allies by expanding system boundaries, but you can also expand system boundaries by reaching out to new partners. The mayor's office can include citizen advisors from every neighborhood, which would provide the mayor with insight into conditions everywhere, not just the pockets of the city she knows best. Corporations can follow the guidance of people who use their products and harvest their raw materials. In so doing, they expand the definition of what's considered important information. That's an expansion of the system's boundaries.

It isn't always the case that boundaries need to be expanded. Sometimes, the healthy choice with boundaries is actually to shrink them.

In ecology there is the concept of refugia.[4] Refugia are small bounded off bits of ecosystems that might be separated from the rest of the ecosystem. These are bits of ecosystems that have some sort of physical separation that allows them to maintain a more stable climate microsystem in the midst of a changing landscape. This separation can protect refugia species from threats such as a novel infectious disease. By having a strong, narrow boundary, refugia stay safe during tumultuous times. They can hedge against disaster, and they can be a contributor to recovery.

Systems that have strong boundaries can also be places for experimentation and learning. Such boundaries can protect tender new experiments from some of the pressures that might be found in larger systems. A new project, including a multisolving project, might need a boundary that frees the team that created it from some of their ordinary duties so they can have time to dream and experiment. An organic farm might need to exist outside of the commodity market in order to thrive. Land might need to be removed from the real estate market and into a community land trust to preserve affordable housing.

There is no "right" size at which to draw a boundary, even in your mind, as long as it protects health and helps the system meet its goals without externalizing harm onto other systems. It may take time and experimentation and conversation with others to determine the best boundary, and it may even need to change over time as the world or your work changes.

Good Systems Bets

In the last section we looked at some elements of a systems stance. Taken together they offer a way of meeting the complexities of systems. They suggest a "way" to act, but they still leave an unanswered question: What to do?

There are as many ways to act in systems as there are systems to act in and people to act. But there are a handful of types of action that tend to be good investments in systems health. Facing tumultuousness, crises, and uncertainty, you want systems at the top of their game. Healthy, resilient, responsive. You want to invest in the health of systems so that they can serve well for the future, even if you aren't fully certain what that future will bring.

If someone told you tomorrow was going to be a big day, what would you do? You don't know if you'll need to run a five-mile race, solve differential equations, or mediate a big argument. What do you do? If it were me, I'd drink plenty of water, eat a good dinner, and get a good night's sleep. This section is about the equivalent investments for complex systems. Even if you don't know where the thresholds are, even if you aren't sure if a new feedback loop is about to kick in, you can still invest in a system's health.

I think of these actions and investments as good bets that will likely be helpful no matter what tomorrow brings. Because they have the potential to improve systems overall, these good bets can often offer ways to multi-solve. One prime example is increasing the equity within systems, which can improve both public health and economic productivity, addressing two significant problems with one intervention.[5]

The following list is a general guide of course, so consider it a starting point. With some thought and experimentation, I hope you will be able to apply these good bets to the systems in which you live and work.

Invest in a system's capacity to cope with change. Bolstering the capacity of a system to cope with change can be a good investment, no matter what the future brings. Think not only natural disasters and economic shocks but also positive changes. Is your team able to handle your company landing a huge new client? If your nonprofit's methodology gets national attention, can you meet demand? Your society (finally) gets serious about building a low-carbon economy and the pace of change is dizzying; are citizens able to adjust to new technologies, fill new types of roles, and meet their basic needs even through changes to infrastructure and incentives? Are people well supported through the waves of change?

What coping capacity should you focus on? To answer that you will need to define the system. Is it a business, community, region, or ecosystem? Where should the system boundaries be drawn? You will also need to identify which types of shocks you think are most likely to occur. Practice scenario thinking. What are the most likely threats? What are some possible opportunities? Some might come from outside the system boundaries, while others might emerge from within as a result of the system's own behavior.

The capacity to cope is a stock (or series of stocks) available to help manage disruption. Depending on the system and shock, they might include financial savings, neighborhood resourcefulness, knowledge of first aid, idle lines in the manufacturing plant that could be pressed into service, soil carbon reserves, dollars available for interest-free loans for adoption of new technologies, and spots in job retraining programs, to name just a few.

The stocks you identify will have inflows that can increase coping capacity, including training, investing, funding, practicing, equipping, and resting. Spending, using up, exhausting, and wearing out are outflows that deplete it. If you intend to increase a system's coping capacity, you can increase the inflows that build it up or slow the outflows that drain it. Or you can do both.

For some systems, you may have the opportunity to restore an innate capacity to cope that has eroded over time. To locate these opportunities, look at the outflow of the coping mechanism. What's depleting it? Can you stop or slow that depletion? Look at the inflow. Is there anything you can do to speed up its replenishment? Think expansively. For instance, there might be one employee at a business who knows how to protect key equipment in a power outage. Turning his knowledge into procedures and checklists that any employee could use expands coping capacity, increasing the stock of people prepared to act in an emergency.

Communities can strengthen their inherent coping capacities too. Some of these are physical. Cooling centers help communities cope with heat waves. If those same cooling centers have energy storage such as batteries to power refrigerators, they can keep medicine cool during power outages. Phone lines and internet cables are stocks that help communities receive and send information in a crisis. Investing in the coping capacity of the system might look like upgrading so that all homes have reliable connectivity. Other coping capacities are not physical. Skills, knowledge, and creativity represent stocks that can be drawn on in times of disruption.

Look to build coping capacity by strengthening connections between the parts of systems. In a well-connected neighborhood people will be able to rely on each other to help during an emergency. People will know who needs help and how to get it to them.

Another way to help systems better cope can be to build in some redundancy. That might be in supply chains, where you forge relationships with a couple of different suppliers, or it might be in job descriptions, where you make sure every critical role on your team can be filled by more than one trained person. Your system can still cope, even if part of it gets disrupted for a while.

Investing in the capacity to cope is a good idea for one final reason. The stock we draw on to cope during moments of disruption or destabilization can also be beneficial when things are steady. Neighborhood cohesion is a source of well-being every day, not just during shocks. Creativity will enhance your team's work daily, not just when an unexpected glitch hits your operation. That's another example of multisolving of course, an intervention that pays off in a crisis but is there to support people during ordinary times as well.

Improve the accuracy and timeliness of flows of information. You can often make a system healthier by improving the accuracy and timeliness of its information flows, no matter what the future holds. Balancing feedback loops can close critical gaps only if they have detected those gaps: distortions that cover up the real state of the system or delays that slow down that information arriving where it could make a difference and prevent balancing loops from working well.

The good news is there are many ways to improve information flows. When data about the state of the system is missing, communities can sometimes even collect that data themselves. For instance, Dr. Sacoby Wilson and his colleagues at the University of Maryland are working with communities to deploy networks of low-cost local air quality sensors.[6] The sensors help gather information about air pollution in real time. The sensor network gives residents access to data about air pollution that they can use to influence local decision-making.

Farther south, in Atlanta, Georgia, scientists, university students, and community-based organizations built an app that community members can use to enter instances of "stressors" to water systems, including standing water, illegal dumping, and ineffective stormwater infrastructure.[7] The maps produced by this community research help prioritize sites for investment and remediation by city government.

It's not only scientific or technical data that can be sources of timely information. Artists also help people recognize important gaps between how we wish systems were and how they actually are. Maya Lin, the artist who created the Vietnam Veterans Memorial sculpture in Washington, DC, created the What Is Missing? project, which she calls her last memorial. It is an installation that "focuses attention on species and places that have gone extinct or will most likely disappear within our lifetime if we do not act to protect them."[8] One part of the project was the relocation of an eerie "ghost forest" of forty-nine dying cedars from the New Jersey coast to New York City's Madison Square Park. The trees were victims of saltwater inundation from climate change.[9] Like output from Dr. Wilson's sensor network or Atlanta's citizen scientist report, the installation is part of a feedback loop, in this case about sea level rise.

Make careful use of energy and materials. Our planet's climate is an example of a system whose inflows and outflows have been thrown out of balance. Inflows of CO_2 to the atmosphere are larger than outflows from the atmosphere to forests and oceans. As the stock of CO_2 rises, the planet warms. To find our way out of converging crises we will need to bring inflows and outflows such as these back into balance with each other.

Balancing inflows and outflows of nonrenewable resources means bringing the flows whereby the economy uses fossil fuels, precious metals, and fossil aquifer water to zero (since the inflow to these is zero). Renewable resources also need inflow/outflow balancing for the long term. Trees grow, fish breed, the sun sends energy our way. Those are inflows. For these resources to be sustainable, the rate of use globally (the outflow) needs to match the inflow. At the same time, the basic needs of billions of people around the world are not being met. Those people will need access to more of the outflow. Taken together, that means many systems, especially in the Global North, will need to draw less from the outflow of renewable resources.

Weaning systems from nonrenewable resources and using renewable resources more frugally are also good bets when facing an era of shocks and tumultuousness, for three reasons. First, as nonrenewable resources are depleted they may become costly and volatile. The more dependent your system is on them, the more vulnerable it will be to shocks and disruptions.

Second, new policies and rules are being created in response to the dangers posed by reliance on nonrenewable resources. In addition, new protections are being advocated to better manage the extraction of renewable resources. Though good for the whole system in the long term, these policies may come as shocks to smaller systems. Steer your art museum over to renewable energy now if you can. You will avoid the financial risks that continued reliance on fossil fuels might bring. Begin learning how to use sustainable materials in your construction company now. You'll be ahead of the curve if your city institutes new requirements.

Third, nonrenewable resources are often (though not always) extracted far from where they are used. The more reliant you are on distant resources, the more your system is exposed to the risks along the supply chain, from conflict to natural disasters.

For all of these reasons, more efficient use of energy and resources is often a multisolving solution. It can help address global crises like biodiversity loss and climate change while making systems healthier and more resilient.

Another way to reduce use of energy and materials is to try to meet nonmaterial needs nonmaterially. Current systems use a lot of energy and materials to try to meet goals that may not truly be satisfiable. Beyond a level where important basic needs are met, additional consumption does not necessarily produce additional happiness or well-being.[10] Humans need love, a sense of belonging, a sense of contribution. We need excitement, challenge, and creativity. These needs can be met with little expense of energy and materials. As Donella Meadows said, we can learn to meet our nonmaterial needs nonmaterially.[11]

Focus on solutions with long lifetimes. Long-lasting infrastructure—buildings, transportation systems, electric power plants—are better than their quick-fix alternatives. If designed well, they will deliver benefits for decades. Large public infrastructure projects, like transportation systems or large tracts of housing, have a multiplicative effect. If they are efficient with materials and resources, they can influence the impacts of thousands of people each day for decades.

Build the power to change systems. Cultivating the power to change systems is another good bet. One way is to build connections, the essence of many multisolving strategies. People and groups can come together to address problems on the strength of common interests and the possibility of simultaneous improvement of many parts of systems. Of course, even after initial problems are addressed, the system will have changed. New connections will exist. A network will have emerged and, with a little tending and attention, can remain, with the power to change systems again in the face of future need or opportunity.

Community organizing brings individuals together to change things none can change alone. Strategic partnerships—like those between labor groups and environmentalists—find ways to align political power and resources to accomplish system changes that both movements care about.

Analysis and a compelling story can be part of building the power to change systems as well. Can you help people understand why some aspects of a system need to change? Can you explain why the system does what it does? Can you explain how a change would lead to better results?

The power to change systems can grow via reinforcing feedback. Small victories can garner attention, which can bring more people and energy, which can increase the capacity to change systems, generating yet more successes.

No matter how you go about it, building the power to change systems is a good bet because it is useful no matter what the future brings. A community that is better able to influence decision-making and spending can both advocate for itself in the face of threats and ensure it's included in new emerging opportunities.

Increase equity. Boosting equity can improve the health of systems, making them more equipped to meet shocks and navigate complexity. That makes investing in equity a good bet.

Inequity creates sacrifice zones inhabited by marginalized people neglected by the decision-making processes of systems. Marginalized people's knowledge, expertise, and interests are left out of decision-making. In other words, equity starves systems of balancing feedback. In so doing it limits their ability to self-regulate.

The people living downwind of refineries know that fossil fuels are dangerous. Had their voices fed into decision-making decades ago, fossil fuel companies would have had more regulations. Clean energy companies would have had more economic advantage. The shift toward clean energy may have happened sooner as a result.

By starving systems of feedback, inequity slows response time. After all, who feels the consequences of a problem in a system first and strongest? Often those who are the most marginalized, with the least impact on decision-making.

As we face converging crises, we need more early warning, not less. We need more feedback loops—more collective intelligence—not less. That makes increasing equity a good bet. Doing so makes systems smarter and more responsive.

There are practical reasons for reducing inequity. There are also ethical ones. You are a unique and special part of the systems you take part in. So is everyone else. Just as unique. Just as special. In interdependent systems, no part is more important than any other part.

There are no one-size-fits-all steps to improve equity, but we'll look into equity in more detail in chapter 10.

Practice solidarity. The final good bet on this list is solidarity. Examples include mutual aid projects in communities and international assistance between countries. Financial support, technical support, and the social safety net are all examples of solidarity. When people and communities support each other in this spirit of solidarity, it allows resources to flow flexibly and dynamically. I may help you today. You may be there for me next year.

I come to solidarity last because it combines some of the other good bets on this list. It's a form of increasing the system's ability to cope, as when volunteers arrive to help after a disaster. Sister cities that support each other's citizens are another example. Solidarity means that we don't all have to always have fully self-sufficient coping capacities. We know that some other systems have got our backs.

Protecting one another's civil and human rights is an act of solidarity that can also make systems more equitable.

Solidarity can connect people and communities together in a way that also builds the power to change systems.

Solidarity belongs on this list for another reason. It can speed up a system's ability to change. We know that being able to change quickly and nimbly is a needed capacity for tumultuous times and converging crises. The architects of the Green New Deal proposal in the United States are well aware of this aspect.[12] The Green New Deal includes provisions for universal education, childcare, and health care. Critics immediately claimed that the proposal went too far. It should limit itself, they said, to climate and clean energy.

Green New Deal thinker Rhiana Gunn-Wright of the Roosevelt Institute has answered this critique many times.[13] She points out that moving much of the workforce from polluting industries to clean ones represents a huge transition. Workers afraid of losing health care in the transition are likely to resist change. Those good new jobs might not appeal to workers who don't have the childcare they need to get training for the new jobs. Workers unsure that they will be included in the new economy are likely to resist change. Having specific plans to care for each other through a big change makes the big change more likely to happen. In other words, solidarity is a lubricant for change.

A few important caveats about the preceding good bets. First, there are rarely single magic solutions that are uniformly appropriate in every situation.

Everything you do comes with the chance of side effects. That's true even for the recommendations in this chapter. You may even have noticed contradictions as you read through the list.

I urge both careful and efficient use of resources and redundancy. How can you have both? Redundancy is inefficient! I don't disagree. Have I mentioned the contradictions inherent in complex systems? Context, judgment, uncertainty, and paradox are part of the terrain. Sorry!

If I were working in a system and trying to find a way out of the contradiction between using resources efficiently and investing in redundancy, here's what I'd do. I'd invest as much as I could in frugality with material resources. I would perhaps push even further than I might have if I weren't also thinking about creating redundancy. I'd use the extra "space" I'd created by my frugal resource use to invest in redundancy. I'd also think hard and carefully about my biggest vulnerabilities. I wouldn't aim for redundancy everywhere in my system; I'd try to target it to my biggest vulnerabilities. Even using that same logic, two people might choose a different path forward for the same system. They might both be right, as much as systems allow there to be "right" answers. Whatever they choose, one hopes they'd approach it with learning attitudes.

Here's a second caveat. Just because a change is good for a system doesn't mean more change in the same direction is necessarily good. Take the first good bet, investing in a system's capacity to cope. If a system already has a great capacity to cope, boosting it might not produce a lot more benefit. Time and resources could be better spent elsewhere. Most of the systems I know and have studied are far from that point. Many have quite limited coping capacity and lack accurate information flows. They are wasteful with energy and materials. Most have at least one form of inequity. Some have many. Most systems I encounter are engineered to be low in redundancy. Most are siloed and disconnected, and many people feel extremely limited in their power to change systems. With this as the general baseline of most systems, all the good bets mentioned in this chapter are worth thinking about. But there may be exceptions. If that's the case for you when it comes to one or more of these strategies, then by all means turn your attention elsewhere and invest time and energy where it seems it will make the biggest difference.

I hope this chapter has given you a taste of some of the ways we can work with systems, even if we can't control them. We started with a systems stance. Seeing systems. Getting comfortable with uncertainty. Treating boundaries and time horizons flexibly and playfully. We then looked at some of the ways we can make systems healthier and more responsive. We explored good bets that are likely to be helpful, no matter what the future might bring. That's a great start for a world of complex systems and a time of rapid change. And that's not even the extent of our options.

We can also act to steer systems. As individuals, we are often limited in our power to address system problems. But by working with others, spanning the borders and boundaries of human-created silos, we can take advantage of the interconnections within systems. We can shift system structures, and we can change some of the underlying beliefs and values that shape those structures. Steering systems—how and why—is the subject of the following chapter.

? Questions for Reflection

- Which of the elements of a systems stance seem hardest or most counterintuitive to you? Pick a length of time (a week, a month) and try putting that element into practice. If you like, keep a journal where you reflect on the experience.
- Which of the elements of a systems stance come easily to you? Do you think there were experiences in your life or education that helped?
- How might you support others you live or work with in practicing the skills of a systems stance? What support might you need from others?
- Pick a system you care about and have some influence within. Which of the good bets listed in this chapter seem most appropriate for that system? Do you see any opportunities to act alone or with others to help bring that possibility into being?

Surprises

Out beyond the edge of prediction,
when you lose the well-marked path
in the tangle of new vines
and the rubble of crumbled
        structures,
you'll need a map.
You'll need a compass.
Look deep,
join hands.
You will find your way.

CHAPTER EIGHT

Steering Systems

Complex systems are always in motion. Multiple flows fill and drain multiple stocks. Feedback loops respond to changes in stocks and send out ripples of change in all directions. Each of us—individually, in teams, in organizations, in businesses, in social movements—exists within this swirling motion. We are often small relative to larger systemic forces. The system is a giant ocean of weather and waves, and we are a tiny speck floating in that ocean. No matter how smart or hard-working we are, we can't always make systems go where we want.

Yet we want to divert systems away from danger—climate change, economic risks, political instability, inequity. We want to create persistent fundamental change. We want systems to seek and find balance. We want essential stocks that are far too small to grow and grow quickly. At the same time there are actors with other ideas for our systems, other goals, other sources of power and flows of information. There are incentives and pressures percolating down from higher levels of nested systems, along with change bubbling up from lower levels.

Emergency-scale shifts in system behavior are required, but we are relatively powerless to make those shifts happen according to any sort of detailed specifications on a timetable of our own choosing. Phew! I don't know about you, but for me that creates some significant tension. We can't control systems, yet our well-being, maybe even our survival, requires many systems to move in different directions. The good news is that if we think differently, if we think about working with systems instead of trying to control them, there are ways to transcend this tension.

Beyond Control, the Possibility of Steering

We can't control systems, but we can work with them and with each other. We can nudge them. We can find windows of opportunity to act. We can change how systems are connected and, in the process, change their behavior. We can change underlying conditions, adjust the rates of flows, and strengthen or weaken feedback loops. We can't always make systems do everything we want, but we can influence them, and I think of this influencing as steering systems.

In fact, we steer systems all the time. Remember that shared heating system whose mismanagement I described in chapter 6? It's been my teacher of the art of managing systems for twenty years. In the early years of our communal heating efforts, we tried to schedule the stoking of the boiler as a group of twenty households. People would forget their shifts, the boiler would cool down, and we had chilly mornings, cold showers, and grumbling. If there was a missed stoking it wasn't always clear whose job it was to run up and add a few more logs to the fire; there would be lots of emails as forty adults tried to communicate to swap shifts and arrange coverage. A few years in, one of my neighbors had an idea: we divided ourselves into "teams," one for each day of the week. It was up to each team to devise a way to ensure that the stoking slots for that day were filled. Now we had a smaller group of people to communicate with and, frankly, a little less confusion to hide behind. People still forgot, but there was a small group of teammates watching out to make sure the day's stoking went well. The system behavior (both the number of missed stokings and the level of grumbling) improved because we changed the structure of the system. We cut some information flows (like grumbling that didn't accomplish much). We strengthened others (if you are cold on a Tuesday, you know to call Tuesday's team). We steered the system.

Long before the COVID-19 pandemic, the organization I cofounded, Climate Interactive, was a virtual organization, with team members scattered around the world; we had to figure out how to make that work well. We were pushing hard toward big goals and moving fast. Like any organization, we had our conflicts and differences of opinion. And sometimes we'd try to work out those conflicts via emphatic email exchanges that rarely resolved much and sometimes escalated into hurt feelings, confusion, or lost productivity. Dismayed by this system behavior, the other cofounder, Andrew Jones, suggested a simple rule: "No emailing if you're mad." If we were upset with a teammate, we had to pick up the phone and have a conversation where we could hear tone of voice and remember our good, hard-working colleague who had the same goals we did. We didn't stop misunderstanding or disagreeing, but with that simple rule we began to strengthen rather than weaken relationships in the process of sorting our differences out. The systems behavior you might call "number of unproductive exchanges per month" improved. Andrew steered the system.

This anecdote illustrates the distinction between control and steering. Notice that Andrew steered with a light hand. If he'd been more control oriented, he might have started monitoring all inter-team emails. He could have penalized people on their annual reviews if they were embroiled in too many email conflagrations. Or he could have spent lots and lots of time refereeing individual disputes. Instead of trying to control individual system elements—his colleagues—he shifted the way information flowed between them.

Once you start to think in terms of releasing control, the possibilities for steering are plentiful. Let's look at some of the ways to steer systems.

Steer with Vision

My mentor Donella Meadows wrote that she never started any project, from a new garden to a new book, without a formal visioning process.[1] Donella's visioning process included finding a quiet place, closing her eyes, and imagining the future. She advised using all of your senses, tasting, feeling, and even, if you can, conjuring up the smell your desired future. Is there fresh bread baking in the oven of your net-zero house? Children's happy voices drifting up the stairs from the cooperative outdoor school in your neighborhood? Peter Senge has a chapter about "shared vision" in his book *The Fifth Discipline*. In it he describes how a shared vision fosters a long-term perspective and gives a group a common point of reference out in the future.[2]

Vision—a well-developed picture of the world you would like to see—offers a way to steer complex systems. The more vivid, detailed, and fully imagined you can make it, the more powerful a steering force it will be. Since systems are multipotential, small changes made at opportune times can trigger large shifts and move systems in new directions. Vision helps you identify which systems variables you hope will move in the direction of your vision.

A formal visioning process changed my life almost thirty years ago. My husband, Phil, and I were newly married and living in an apartment near Boston while I finished graduate school. A friend led us in an exercise to imagine the future we wanted together. We saw ourselves living on a farm, working with others, learning and practicing sustainability and organic agriculture. We saw ourselves teaching classes, growing food, and raising our kids with lots of other kids nearby. Our imagined family spent a lot of time outside in nature.

At the time we undertook the visioning exercise, our lives didn't look like the vision at all. We were living in the city. We had a tiny community garden plot. Phil worked in biotech, and I was studying neurobiology. We had no idea how we were going to get from the lives we had to the ones that seemed to be calling us.

A few years went by. Then one day we got a phone call from our friend Donella Meadows. She was full of excitement about starting a new project that would include cooperative living, an organic farm, and a research institute. She thought we might be interested in it. Also, she had a small grant to begin the work of the institute and wondered if we were available for hire.

Although Donella had little funding, and we had a months-old baby at the time, we said yes. Both financially and career-wise, the leap was a huge risk. Looking back at those young parents, I marvel at their courage. I also see how much that visioning process fortified our courage.

The famous saying "chance favors the prepared mind" is a tip of the hat to vision. If a sudden moment of opportunity arises, you are more likely to recognize it and step into it. Without the vision we'd still have gotten that phone call. But I don't think it would have been as salient; it would have been just one more open question amid the complexity of life. And we might not have noticed how well that opportunity fit with the direction we wanted to move.

Almost thirty years later, we still live in the community that we leapt into all those years ago. We worked at the institute for a decade, and my current work is a direct offshoot of those years. As we'd envisioned, we raised our children as part of the community, growing a lot of our own food, with a farm and a forest as part of their education. One vision, and one invitation to move toward vision, changed so much about our lives.

Years after my own vision-inspired leap, I taught in a program based on the life and work of Donella Meadows. One of the skills that we taught was visioning. Each participant in the program worked on articulating a vision for themselves and their work. More than ten years later, many visions put forth in those years have come to fruition.

One participant in the program, Any Sulistyowati, envisioned an eco-village in her home of Indonesia, constructed from local materials using sustainable building techniques. She pictured bountiful gardens and vibrant community. When Any first imagined that eco-village, she was sitting in Vermont, imagining something yet to be created on the other side of the planet. More than that, she had no land, no building materials, and no fellow community members. Yet I remember how vivid her descriptions were and the sketches she painted with words and pictures when she talked of her vision.

Not long ago, more than a decade after she first envisioned the eco-village, I talked with Any via Zoom to catch up with each other's lives and work.

She was sitting under a steeply pitched roof of wooden beams and smiling hugely. "Do you know where I am? In the house that I envisioned." She talked of the views from different windows, and of gardens and how she shares a piece of land with the nonprofit training center she founded.

How much of what you have created is like your vision, I asked. How much is different?

Any explained that out one window there's a view she never expected. The house looks down on the lights of the city. In her initial vision, she thought they would find land farther from the city. And she said, "While we have a village, it's not the one I imagined, the intentional community that people would move to. Instead, we found land on the edge of a village, and that village has become part of our community; many of the children from the village participate in our learning center."

She explained the circumstances that led to this affordable piece of land and how the vision changed to accommodate the options before her. And she hasn't fully given up on her initial vision. There's another piece of land, farther from the city. She thinks one day a bigger community could be created there.

Any's story beautifully illustrates one thing I've found to be true of visions. They have a way of coming to fruition. But they also bring surprises.

Cultivating vision isn't easy. It creates tension between the world as it is and the world you want. It may feel tempting to lower your vision to reduce that uncomfortable tension. Don't succumb to that temptation! Envision what you really want, not just what you are willing to settle for. Paint it in detail in your mind. Come back to it and add to it. Write about it or draw or paint it. Make it as vivid and real as you can. Carry it with you. I feel certain that eventually, whether right away or after many years, you will see opportunities. Along the way, your vision will give you courage and clarity.

Pursuing your vision will also have ripple effects. Any's vision has changed the lives of both her neighbors in her community and the students who pass through to learn. Can you picture the health, sustainability, and equity that could result if more and more of us dared to vision and then acted on those visions?

Vision lies at the heart of multisolving. In fact, multisolving starts with a vision that involves several elements of the world improving at the same time.

A place for Any's family to have healthy food and live in a sustainable way *and* a sense of community *and* a center for education. Cleaner air *and* a stable climate. Greener cities *and* cooler homes *and* less flooding from stormwater.

Vision can encompass as many facets as you are willing to imagine.

Steer with Values

Systems reflect the values of the people who inhabit them. A system shaped by people who prioritize efficiency will put people and nature in harm's way if doing so serves the cause of efficiency. A system created by people who value the well-being of some groups over others will be full of disparities and inequities. A system that values next year's financial metrics will bias its decision-making toward short-term profit. On the other hand, a system that values equity and justice will make sure that everyone has enough. A system that values stewardship will account for the needs of nature and future generations.

Rules, laws, incentives, investments, and policies all shape how systems operate. They control the rates of flows that in turn change the level of stocks. They influence the strength of feedback loops. And rules, laws, and incentives are shaped by values.

When values change, policies, laws, and incentives change, systems shift. If you can change the shared values in a system, you can steer that system. Of course, it is easier said than done to change the values of a large system like a society. Still, we can choose and control our own values, and they can guide our actions within systems. The best course of action isn't always obvious at a given moment, but values can provide clarity for our actions. And through our actions, our values extend outward to influence systems.

Values, like vision, steer systems in moments that you can't always predict. You may have partial information or feel overloaded by stress or confusion. Values help you know which direction to move, even when the situation is confusing, complex, and overwhelming. But values can't guide you if you are not clear what they are. Moments of danger or opportunity may not be the best time to slow down and grapple with your values. Instead, whether alone or as part of a collaborative, it is better to be clear about your values well in advance of potential moments to steer systems.

So take some time to grapple with your values. What are they? Where did they come from? How have they changed over time? Are the crises of this moment challenging or changing any of them? Which values have you chosen, and which have you absorbed, osmosis-like, from cultural institutions or your family? Which values do you deeply wish to hold on to, and which ones can you let go?

Acting on values that are different from those of the dominant society isn't always easy. It can feel unnatural. It may come with costs, from social friction to lost business opportunities. On the other hand, the more explicit you are about your values, and the more often you make choices within systems accordingly, the more the systems you move within will reflect your values. Can we draft this corporate policy to embody our belief in the inherent value of all people? Can these architectural plans reflect our commitment to ecological well-being? Imagine living and working with systems fully aligned with your values. That sounds pretty joyful to me!

Sometimes closely held values seem to contradict. Say your values include frugality and sustainability. What should you do when the sustainable option is also more expensive? Though I sometimes feel frustrated when I am caught between two important values, I try to take that "trapped between my values" feeling as a call to deeper reflection.

What are the true units of frugality? Are they measured in dollars, or something else like happiness per unit of time or natural resources? Maybe the sustainable option is expensive but maybe it's also frugal in a deeper sense of frugality? Or maybe the need can be met without a purchase at all through sharing or mending? Then you've found a path that is frugal and sustainable!

While sometimes values really do sit in conflict, I've been surprised at how sometimes a deeper look finds a resolution.

Steer with Simple Rules

Simple rules are the organizing principles that give rise to self-similar patterns in systems. The same simple rule can influence different domains within a system. At work, my colleagues and I follow an unwritten simple rule, "get a second set of eyes." If I'm writing an essay, I'll ask for a colleague to review it for typos and the logic of my argument. If someone on the modeling team writes new equations, she'll ask a colleague to review them. Someone else might run a grant proposal past colleagues before it goes out the door. From writing to analysis to fundraising, the same simple rule guides what we do. Without it, our system would be different; we'd probably produce lower-quality work. Because one simple rule can reach into so many different domains, simple rules offer a good way to steer systems. Another company might have a rule to "get four sets of eyes on it." They might make even fewer mistakes than we do, but they'd probably be slower too.

Carolyn Raffensperger, executive director of the Science and Environmental Health Network, is a fan of simple rules. She's helped popularize an important one: the precautionary principle mentioned in the previous chapter. The precautionary principle can be applied to many types of new technologies. It's flexible enough to guide action in quite disparate domains. Precaution makes as much sense for genetically modified crops as it does for new chemical compounds, nanoparticles, or artificial intelligence.

Trained as both a lawyer and a scientist, Carolyn also sees room for the logic of simple rules in law. She has proposed the idea of supplementing the legal "reasonable person" standard with a "respectful person" standard in a variety of situations: respect for the rights of women, for the rights of elders, for residents living near industrial sites, perhaps even respect for the rights of nature.[3] If such a simple rule were to be accepted across many branches of law, think of the changes that would result across society.

If you'd like to bring simple rules into your work, I recommend the Human Systems Dynamics Institute's guidelines.[4] They suggest that simple rules be phrased in the positive (what we do, not what we avoid). Simple rules should be active. In fact, the experts at HSD suggest that the first word of a simple rule be a verb.

The first challenge is to identify what simple rules are currently operating in a system. Most of the simple rules that guide our actions in systems are implicit. If you want to change simple rules to steer systems in a particular direction, you first need to become aware of them.

The second challenge is to put a new simple rule in place. For example, during the last few years, I have wanted my project work to go deeper and be more impactful. But sometimes days and weeks would go by where I could barely keep my head above water. I'd spend large chunks of my days answering emails and doing small projects and never seem to have any time left over for sustained thought. I eventually realized that I was operating, albeit subconsciously, from a simple rule: "respond to everyone who asks."

I chose a new simple rule instead: "protect some time for deep work." That simple rule led to some changes. I started using a new system to block out my workdays, scheduling time in the calendar for writing and thinking as well as for meetings and phone calls. I found a place without phone or internet access that I could use as a writing office and scheduled some writing retreats.

My time spent writing trended upward. That's because the structure of the system shifted. The stock of hours for deep work increased because I found ways to stanch the outflow from my stock of hours in the workweek. Other aspects of the system changed too. I had to ask for more help from colleagues, and I spent less time on social media.

Applying a new simple rule, one I chose intentionally, steered the system of my professional life. Of course, it wasn't a seamless or instant transformation. I occasionally forgot my new rule and slipped into old habits. Some weeks I applied it consistently, some weeks spottily. But over time following the simple rule became more comfortable. And when I began to see my productivity increase, it became easier and easier to justify the change.

Changes to simple rules can help solve multiple problems at the same time, so they, too, are a tool for multisolving. "Measure the well-being of people and nature" is a simple rule. If an index of well-being was followed throughout society instead of the Dow Jones index, newspapers and websites might feature dashboards showing daily numbers of hungry children, maternal deaths, species extinctions, literacy rates, women's pay rates, and CO_2 concentrations. I'd like to think that using dashboards like that would lead to many of those indicators moving in directions we'd like to see. One new simple rule could create the conditions for progress on multiple problems at the same time.

Steering with Vision, Values, and Simple Rules in Combination

In 2007 the small town of Greensburg, Kansas, was leveled by a tornado. A dozen people from the town of 1,400 died. Many public buildings and private homes were damaged or destroyed. The leaders and residents were faced with a massive rebuilding effort. That process of rebuilding, as documented by the *Washington Post*, was steered by vision, values, and simple rules: "Just a few days after the tornado, Steve Hewitt, the city manager at the time, said that he sat down with John Janssen, then the city council president, and other leaders in the City Hall parking lot to decide the future of Greensburg. They quickly reached a conclusion: They should rebuild with smarter, energy efficient buildings."[5]

State and federal funds and new investment flowed into the rebuilding process. Hundreds of decisions had to be made. Building codes, design and building firms to hire, materials to be used, and much more. The values of efficiency and stewardship, the vision of a green economy, and simple rules like green building standards steered those choices and shaped how the town reemerged. With an anchor of vision, values, and simple rules clear, each decision didn't have to be debated or thought through in isolation. The system had a sense of direction.

Just how much the system had been steered is clear in how different the town is today from a typical town in the United States. According to that same *Washington Post* article, a decade after the tornado struck, "Greensburg draws 100 percent of its electricity from a wind farm, making it one of a handful of cities in the United States to be powered solely by renewable energy. It now has an energy-efficient school, a medical center, city hall, library and commons, museum and other buildings that save more than $200,000 a year in fuel and electricity costs, according to one federal estimate. The city saves thousands of gallons of water with low-flow toilets and drought-resistance landscaping and, in the evening, its streets glow from LED lighting."

Not only did the town steer with vision, values, and simple rules, it multisolved as well, saving money, saving energy, and protecting the climate.

Scaling and Spreading

The examples of steering in the last section mostly concerned small systems. But what about larger systems? Vision, values, and simple rules can help steer them too, for three reasons.

The first reason is that influential leaders at high levels in large systems also act out vision, values, and simple rules. The CEO articulating a vision of environmental sustainability can influence the whole company: the R&D department could be directed toward efficient use of materials; investments in the physical plant could prioritize efficiency; company communications could highlight the commitment to sustainability. Billions of dollars of value and spending can be redirected with a shift in the vision of just one leader.

Second, organized groups of citizens, voters, stockholders, or employees can also push leaders to adopt new visions, values, or simple rules. Movement building (reinforcing feedback in action!) generates the power to shift visions, values, and simple rules in large systems. Ordinary people might not immediately influence decision-makers at higher levels, but by building collective power together, people can change these three powerful forces that steer systems.

Third, new or different visions, values, and simple rules can scale and spread from lower levels of systems to higher ones and from one domain of a system to another. Steering small systems can sometimes steer larger ones! To understand this sort of scaling and spreading, we need to explore another property of systems: coherence.

A system is coherent when similar organizing principles are used at different scales or in different domains. Figure 8-1 shows a fern as an example of a coherent pattern. Fern fronds are composed of smaller structures called pinnae. In turn pinnae are composed of pinnules, which look like "bumps" on the pinnae. Do you see how, from pinnules to fronds, the same shapes are repeated? That self-similarity is coherence. Frond to frond the pattern is self-similar: that's coherence across domains but at the same scale. But notice also how the pinnae look like miniature versions of the frond. That's coherence across scales.

Branching patterns in trees, river deltas, capillaries, and neurons also display coherent patterns. Have you ever studied the silhouette of a tree against the sky? The branches have branches, which have branches, which have twigs. Next time you have a chance, look closely at the branching pattern. No matter what scale, tiny twig or giant branch, the angle will be self-similar. That's coherence.

Every tree species has its own characteristic silhouette. Each is internally coherent, but they aren't as coherent across species. The angles of the branches (the organizing principle, a simple rule) are a little different, from tree to tree, as are the lengths branches grow before they sprout new branches.

While self-similar organizing principles in ferns and trees relate to angles, shapes, and growth habits, in other systems, visions, values, and simple rules can provide self-similarity. The map in figure 8-2 shows an example of coherent visions. "Healthy streets" is a phrase used to represent a particular pattern of urban development. A city with healthy streets is one that provides many options for residents to walk and cycle. It is rich in both green space and public gathering space. Housing is integrated with shopping, schools, hospitals, and jobs. A city might develop a vision of itself moving in the direction of healthy streets. Year after year, policies, planning decisions, and investment flows would be influenced by that vision, shaping systemic structure from municipal bonds to bike path routes. Over time, amenities associated with healthy streets would become more and more common.

If additional cities each subscribed to that vision, they would make similar choices and develop in similar ways. At least when it came to transportation and design, they would be coherent across domains, similar to the pinnae of similar sizes on different fern plants.

Coherence across scales would happen if city, state, and federal transportation decision-makers were all aligned with a vision of healthy streets. Similar visions would be embodied at different levels of the system.

Figure 8-1. A fern is an example of a system with a high level of coherence, producing self-similarity from frond to frond and between fronds and pinnae. *Illustration: Molly Schafer.*

Values and simple rules, like visions, can be coherent across scales and domains, steering systems and their behavior similarly in different geographic locales, different industrial sectors, and at different scales.

Now that we have a sense of what coherence is, we can begin to explore another way that change ripples through systems. How do visions, values, and simple rules spread and scale? Sticking with the healthy streets vision, how might the vision first cultivated in one city spread to other places? Maybe the director of the transportation department in one city speaks at a national conference, presenting a vivid picture of that city's vision and some of the early results. Visions are inspiring and infectious. Transportation

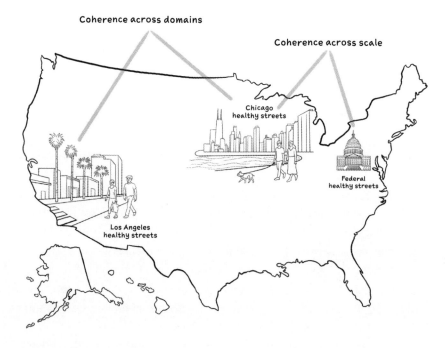

Figure 8-2. Self-similar patterns show coherence across domains, like when two cities adhere to simple rules, like the principles for creating healthy streets, and across scales, like when federal policy aligns with city-level healthy street initiatives. *Illustration: Molly Schafer.*

department heads from other cities might then carry seeds of the vision back to their cities. As the vision takes hold in other cities, systemic structures, policies, and flows of investment will begin to change.

Perhaps a transportation department executive from one city finds a new job in state government, bringing the vision of healthy streets with her. Now the vision has spread to a higher level in the nested system. Policy changes at the state level could ripple back downward as new incentives and supports bring the possibility of healthy streets to other cities and towns within that state.

As the vision gains coherence, its power could grow as well. Officials in several cities could attend trainings together or share the results of pilot projects. Alignment of city and state incentives could speed up progress at both levels. Not only do the local officials have a vision, but now the state is also offering technical support and financing. Through pathways like these, a vision that was once isolated in just a few places within a system can spread throughout it.

A vision that might once have felt small and localized can have far-reaching effects. The scale of the initial vision is less important than its vividness and contagiousness. The same is true for simple rules and values. They can be practiced at a small scale and still spread either laterally to similar systems at the same scale or upward to bigger and bigger systems.

Picture a grassroots organization that focuses on urban orchards. Their meetings generate enthusiastic participation; the community takes ownership of the work. If the orchard group discerns their simple rules for community engagement, those rules could spread to urban orchard projects in other cities. But they could also help bicycle repair organizations or solar energy projects. Organizing principles developed in one context can be transferred to another. Innovations in one part of a system can be replicable elsewhere.

When time is of the essence, we need innovations to diffuse from small pilots to widespread significance. Coherence helps expand the pool of innovators and accelerate change.

Values, vision, and simple rules can also help people improvise, which is another kind of steering. In times of rapid change, given the unpredictability of systems, we don't have a playbook for every situation we encounter. But people who are clear about vision, values, and simple rules in one context can draw on them in others.

Can a team write a grant proposal in twenty-four hours? Well, if they've been exploring their shared values for the last six months, they have a better chance. A neighborhood has to be evacuated due to a flood? If the community's values of mutual assistance are clear, that should streamline the sheltering process.

Simple rules give people the chance to practice new ways of operating when stakes are low. That practice is a resource that can be drawn on in a novel situation. Say a young person is member of a worker co-op grocery store that makes decisions by consensus. Day in and day out, she partakes in decisions about production, marketing, working conditions, and compensation. She becomes familiar with a set of (possibly unwritten) simple rules for group decision-making. If destabilization strikes her community, those simple rules are a seed for improvisation. She could draw on them to help people in her town make decisions about bridge repair or medical supplies.

What's interesting is that the worker co-op does not need to incorporate community disaster resilience into its mission. It can stick to its mission of being a grocery store that operates collaboratively. With high fidelity to the vision, values, and simple rules of cooperativity, the store is establishing patterns that can transfer, in a moment of need, into other systems.

———∞———

Systems are always remaking themselves. Bodies replace cells, forests replace trees, and cities replace buildings and roadways. Regeneration may always be happening but, depending on the system, its pace is not always consistent. In our current moment many people are accelerating the remaking of systems: retrofitting homes for better energy efficiency; deploying new technologies to reduce greenhouse gas emissions; rebuilding, and in some cases redesigning, communities in the wake of more frequent and intense natural disasters.

Vision, values, and simple rules become even more important when systems are being rapidly remade. As trillions of dollars move through the global economy to respond to climate crises, pandemics, and economic shocks, a multitude of systemic changes occur. Some are large, some are small. How do those trillions flow? What do they create? What is left behind? The vision, values, and simple rules that are held by the people embedded within the system steer and direct the change.

What values guide the hand of the transportation planner who sketches out a new bus line? How committed is that planner to undoing historical injustices or keeping children safe on streets? What visions do community members hold for the future of the streets the bus will run down? Can they already see pocket parks and green infrastructure? What simple rules have become second nature? Are there always trees and gardens planted at bus stops? Does every new bit of road construction include creating cycling lanes at the same time? When the new bus line funding deploys, those visions, values, and simple rules will steer and shape how the system changes.

Most of us play small roles within systems, and those systems are often themselves small compared to the systems within which they are nested. Who are we, small and insignificant as we are, to presume to steer systems? Maybe mayors, heads of corporations, or presidents of philanthropies can steer systems, but not us. At least that's how it can feel. Interestingly, I've found that when I talk to mayors, heads of corporations, and philanthropists, they also feel small and insignificant in the face of large and complex systems.

Vision, values, and simple rules challenge this presumption of insignificance. We can actively become clear about our visions, our values, and the simple rules we and our organizations and collaborations embody. We can examine them and refine them over time. We can talk about them with others and build shared visions, shared values, and commitment to a shared set of simple rules. We can act where we find ourselves, knowing that, if conditions are ripe, those visions, values, and simple rules can scale and spread.

The best way to be sure you don't contribute to change is to assume that the systems you live and work within are just too small and insignificant to matter. If you assume that steering only happens at the highest levels, and is only orchestrated by those in positions of extreme power, then you might spend your whole life waiting for your chance to access that power or those levels of influence.

The truth is, the biggest—and in fact the only—influence you have is where you are today. So, if steering systems is your goal, here is my advice: work at whatever scale you're at, wherever you are, but work with steering always in mind, and with high fidelity to the vision, values, and simple rules that make sense in an interconnected and interdependent world.

The Power of Worldviews

In her essay "Places to Intervene in a System," Donella Meadows wrote that changing worldviews is one of the most powerful ways to change a system's behavior.[6] Meadows often used the word *paradigm*, but she used *mindsets* and *worldviews* as synonyms. She said, "The shared idea in the minds of society, the great big unstated assumptions—unstated because unnecessary to state; everyone already knows them—constitute that society's paradigm, or deepest set of beliefs about how the world works."

These deeply held unstated beliefs are so powerful because they influence so much about systems. We design systems and operate within them based on how we understand the world to work.

What a society believes shapes the world they build, and then the world they've built shapes what they believe. To question a worldview is to question everything about the system. And moving through daily life reinforces everything about the worldview. This makes worldviews stable and slow to change.

Nonetheless, a worldview that is out of sync with the world will, if left uncorrected, produce crises. Values, visions, simple rules, not to mention incentives and feedback loops, all arise from worldviews. Systems that embody visions, values, simple rules, and systemic structures derived from a worldview that doesn't fit the world may function to a degree, for a time, but they will also produce harmful consequences.

Since the systems we live and work within are producing climate crisis, inequity, biodiversity loss, and threats to other aspects of well-being, it's worth looking a little more deeply into worldviews.

Two Worldviews

In subsequent sections, we'll explore the dynamics of how worldviews shift. But before we do that, let's look at two important contemporary worldviews, one dominant, and one less so. The future will depend on these two worldviews and how the conflict between them plays out. I think you'll recognize both of them. These labels are generalizations, meant to help us aggregate a number of mutually supporting beliefs and assumptions.

The first worldview, which predominates in much of the cultures, power structures, institutions, and economies of the world today, assumes that the world is—to borrow a phrase from eco-theologian Thomas Berry—a "collection of objects" with limited influence on one another.[7] It is the worldview that underlies colonization and empire, two forces that have contributed to its spread across much of the planet. Core assumptions and beliefs of the Collection of Objects worldview include:

- The world is a collection of separate objects.
- Safety comes from domination.

- Events are explained by direct linear causation.
- The world is organized as a hierarchy with able-bodied, straight, rich, white men at the top, white women below them, people of color still further below, and nonhuman species even further below that.
- Power is the ability to exert force over others.
- Power comes from one's place in the hierarchy.
- Problems can be solved by breaking them down into simpler parts.
- Boundaries are firm and rigid.

We recognize this basic worldview in different domains as white supremacy, patriarchy, and extractive economics. Much of the way the mainstream economy operates under capitalism also embodies this Collection of Objects view. When we look back at the last five-hundred-plus years of history, we can see the signature of this worldview everywhere. From witch-burnings to chattel slavery to child labor to genocide and land grabs from Indigenous peoples, a worldview that says the world is just a collection of objects from which chosen dominators can take and use what they want at will has brought the world much trauma and deep ecological crisis.

The Collection of Objects worldview generally does not consider the behavior and structure of complex systems, believing causation to be short and direct, and giving the most attention to influences that are both recent and close at hand. Worried about crime? Lock people up. Migrants straining resources? Build a wall. Need to feed more people? Add synthetic fertilizer.

Seeing the world not as systems but as separable and discrete entities, this worldview often leads to suppressing and ignoring balancing feedback. The whole point of existing at the top of the hierarchy, which is very important to this worldview, is to not have to consider the needs of anyone "lower down."

I don't know if the majority of the world's people subscribe to this worldview, but aspects of it are common and powerful in these times. And often it is strongest closest to centers of power. And because of the way that worldviews influence systems, the Collection of Objects worldview exists in the structure of our built environment. As an undergraduate in the 1980s I quickly discovered you couldn't find a women's bathroom on the first floor of most classroom buildings. When my college admitted women students a few years earlier they just changed the signs on the bathrooms in the basement and on the second floor to read "women." That's a subtle incarnation of the Collection of Objects worldview, shaping a building's interior and a student's sense of welcome (or unwelcome!). Architect and regional planner Carl Anthony wrote in his book *The Earth, the City, and the Hidden Narrative of Race*, "The dynamics of white supremacy have been clearly reflected in the built environment across generations."[8]

As powerful and omnipresent as the Collection of Objects worldview is, however, there have always been groups, cultures, and knowledge systems living and operating differently. Prominent among them is one I will call the "Web of Relationships" worldview. Some of the assumptions and beliefs of the Web of Relationships worldview include:

- The world is organized as a complex web of relationships in all directions.
- Safety comes from partnership.
- Events are explained by complex causation, including feedbacks and branching chains of cause and effect.
- Power comes from knowledge, skills, and relationships.
- Power is the ability to motivate or activate others with their consent.
- Problems can best be solved by considering the whole.
- Boundaries are fluid and permeable.

Worldviews like this can be found around the world and for centuries, of course. Although there is no single, monolithic "Indigenous perspective," Indigenous thinkers, teachers, and writers from many tribal nations in North America often describe elements of the Web of Relationships worldview when describing their cosmology, thought, and practices. A few examples can help give a sense of a few ways in which a Web of Relationships world-view has been articulated.

Tewa author and professor Gregory Cajete describes in his book *Native Science: Natural Laws of Interdependence* what he sees as a Native American philosophy of science: "Everything is related, that is connected in dynamic, interactive, and mutually reciprocal relationships."[9] Scholar and philosopher Kyle Whyte, an enrolled member of the Citizen Potawatomi Nation, writes in his 2018 article "Critical Investigations of Resilience," "These heritages and traditions, which continue to be tied to Indigenous peoples' current practices and identities, treat moral relationships as complex systems working to promote adaptive capacity, not stagnancy."[10]

Buddhist scholar Joanna Macy points out the many parallels between systems theory and Buddhism in her book *Mutual Causality in Buddhism and General Systems Theory*.[11] The Buddha, she says, "presented causality not as function of power inherent in an agent, but as a function of relationship—of the interaction, of multiple factors where cause and effect cannot be categorically isolated or traced unidirectionally."

Geographer Dr. Farhana Sultana writes and speaks about a "crisis of climate coloniality" and points out that worldviews that have contributed to the current crises can't be expected to solve them. Instead, Sultana reminds us, other worldviews, or epistemologies as she calls them, "emerge from lived experiences that were/are devalued in Eurocentric modernity and climate coloniality."[12] She calls for "real engagement with Global South, Indigenous, decolonial and feminist scholarship, activism and perspectives to be integrated and centered to support revolutionary potentialities and justice as possible pathways forward."[13]

Scientific understanding, from ecology to physics to the theory of complex systems, also aligns more with the elements of the Web of Relationships worldview. Certainly, many of the ideas that we've explored in this book fit better into a world that is understood through a lens that has room for feedback, emergence, thresholds, and nonlinearities, one that tends to see wholes as well as parts and is sensitive not just to proximal cause-and-effect relationships but to distant influences in time and space.

This worldview undergirds systems of gender and racial equity. It highlights the importance of listening for balancing feedback and asking for consent. It expresses itself in economic and other relations of reciprocity as opposed to extraction.

Because worldviews live so much beneath the surface, because they are so often implicit and unconscious, it can be difficult to articulate them. Worldviews that have existed for hundreds of years across spans of history, changes in cultures, and changes in technology can never be summarized by just a handful of bullet points. Of course, there are nuances and variations and subtle changes. I'm sure that others would articulate these two in slightly different ways and point out slightly different distinctions between them. In fact, thinkers, including many who have influenced my own thinking, from Macy and Berry to Riane Eisler and George Lakoff, have long grappled with these two worldviews. Still, I think it's worthwhile to compare these two starkly different worldviews here, because one of them seems to contribute to many converging crises while the other seems to offer valuable solutions.

There are worldviews and then there is the actual world as governed by physics, ecology, and planetary chemistry. Worldviews are always approximations. They are useful simplifications that allow people and communities to function. They provide a working understanding of how to operate, how to survive, and how to thrive.

In that sense worldviews should be measured by their usefulness. Does the given worldview give rise to a set of systems that behave in ways that promote surviving and thriving? If yes, that's a useful worldview.

It's not as simple as that, though. For many of the reasons that we have already explored, feedback on exactly how well a worldview fits the world can be delayed. Really delayed. What if it takes centuries from initial ideas of extraction and domination and empire to feel the consequences in climate change, ocean dead zones, and biodiversity loss? Worldviews, even those that don't fit the world very well, can persist for quite a long time.

Not all feedback from a worldview out of step with the world is delayed, of course. For the people and places not considered to matter in the Collection of Objects worldview, the feedback has been rolling in for centuries. If you and your children were treated as commodities in the transatlantic slave trade, that worldview was clearly a threat. If your lands were stolen or the species on which your culture depended were hunted into extinction by settlers who saw you and your homelands as just a collection of objects, that worldview wasn't matching the world.

But for others, especially those who found themselves at the pinnacle of the hierarchy imagined by the Collection of Objects worldview, things may have seemed to be working, even working well. For some, it may still feel that way today. The more isolated you are from consequences of an ill-fitting worldview, the harder it is to see its flaws. A well-developed sense of empathy can help, of course, and we shouldn't need to experience harm from systems firsthand to understand those harms. Still, one feature of worldviews is that the people with the highest confidence in them are usually those who benefit from them the most.

With global emergencies—from climate change to the biodiversity crisis—escalating, perhaps we are reaching closer to a moment of reckoning with a worldview that has dominated so much of Earth for centuries. Could the Collection of Objects worldview be teetering under the weight of these crises?

What happens when a worldview no longer seems to fit the world very well? What happens when large numbers of people lose confidence in it?

How do worldviews fall from dominance? Is there anything people can do to hasten that process?

The Dynamics of Worldviews

As a philosopher of science and author of the 1962 book *The Structure of Scientific Revolutions*, Thomas Kuhn studied the dynamics of how new theories replace old ones in the scientific community.[14] Kuhn observed that throughout history there were long periods of what he called "normal science." During such periods, which can last for hundreds of years, scientists are in agreement about how some part of nature or the universe operates. For example, there was a time when most scientists agreed that the planets rotate around Earth.

Kuhn called these theories with widespread acceptance "paradigms," a word he used in much the way that we have been using *worldview*. Kuhn's description of how paradigms in science change influenced my own thinking about how worldviews change. During periods of normal science, researchers add details to the body of knowledge, working to fill in some of the specifics of the dominant paradigm of the time. But paradigms, like worldviews, are only human guesses about how the universe works. Eventually a scientist conducting "normal science" will find results that don't fit the paradigm. Kuhn called these results "anomalies."

At first, when researchers discover anomalies, most of the scientific community, sometimes including the researchers themselves, reject the findings. They assume that there's been a mistake with the experiment or with the equipment. When confidence in a paradigm is high, it is tempting to reject evidence that doesn't fit the paradigm instead of rejecting the paradigm! But when enough anomalies accumulate, changes in paradigms do happen. Kuhn called these scientific revolutions.

Kuhn found that there are two driving processes that decrease confidence in a previously popular worldview and increase confidence in a different one. Both are driven by reinforcing feedback loops, which amplify small changes in confidence in a paradigm. According to Kuhn's observations, paradigm changes in science start small and snowball, producing what feels like sudden and surprising change.

The first of Kuhn's two reinforcing feedback loops drives a drop in confidence in the once-dominant paradigm. As confidence falls, people are more likely to accept evidence of anomalies. As accepted anomalies accumulate, confidence in the paradigm falls further. As confidence in the paradigm falls, researchers begin to look for additional evidence of anomalies. When they find more anomalies, they publicize them. As more anomalies are brought to light, confidence in the paradigm falls even further. That leads to more anomalies being discovered. Do you see the reinforcing feedback loop?

Now consider what this could mean for the worldview that we've been calling Collection of Objects. What are some anomalies in this context? What serves as evidence that this worldview doesn't really explain the world?

Many of the impacts of the converging crises we are experiencing are in fact anomalies. If the world really were a collection of objects with white men at the top of some theoretical pyramid, and safety truly were to be found through domination and control, then why is our world in such pain and turmoil? Why is the very basis of our life support system threatened?

Remember that when confidence in a worldview is high, anomalies tend to be glossed over. But if a worldview really is out of step with how the world operates, anomalies will inevitably arise and over time become more pronounced and thus harder to ignore.

In this sense, climate change, rising seas, stronger fires, and more frequent heat waves are all anomalies. So are biodiversity loss, soil erosion, and conflicts over resources and water. A tenth of a degree of temperature increase or the extinction of a few species might not shake most people's confidence in the only worldview they have ever known. But what about five-tenths of a degree or a more severe biodiversity crisis?

If the world really does work the way the Collection of Objects worldview suggests, we should be thriving, yet anomalies continue to accumulate. Are we seeing corresponding signs of a reinforcing feedback loop of declining confidence in the Collection of Objects paradigm?

It's not easy to answer that question, of course, but I think there are some signs. For instance, surveys show that for the first time ever young people are not confident in capitalism.[15] More and more societies are also (slowly) questioning inequities in civil and human rights of women, people of color, and LGBTQ+ people. Each of these questions is a decline in confidence in the Collection of Objects worldview. Activists and advocates are suggesting withdrawal of support from key structures that are aligned with the Collection of Objects worldview and investment in webs of relationship. We hear calls for divestment from fossil fuel stocks and investment in clean energy. We hear proposals to abolish the prison-industrial complex and invest instead in community conflict transformation.

Behind paradigm change Kuhn saw another reinforcing feedback loop at work. At the same time that a once-dominant paradigm is falling, others are rising and competing for greater prominence. In science these are different formulations about how the world works. They might be new, but they also might be ideas that have been around for a while but never became fully mainstream. Kuhn found that those paradigms that successfully replaced older dominant ones did so by virtue of their ability to, as he described it, "solve problems." If a new way of looking at the world helped explain one or more of the troubling anomalies, that solved the problem for the scientists. An increase in the number of solved problems creates increased confidence in a paradigm.

Rising confidence in an emerging paradigm will be accompanied by even more attempts to resolve anomalies. If the emerging paradigm does a good job explaining the world, more solutions will be found. As more problems are solved, confidence in the emerging paradigm increases even more—a reinforcing feedback loop. For the scientists Kuhn studied, solving problems corresponded with conducting experiments whose results were predicted by the emerging paradigm.

What does this process look like in the context of the worldviews at play in our societies? Think of the Web of Relationships worldview as a different paradigm competing for ascendance with the Collection of Objects worldview. Instead of experiments in a lab, we have groups of people seeing the world as interconnected and acting accordingly. People are already, in this moment, solving problems via collaboration, the power of interconnection, and practicing mutual support. Every time such work helps people and communities survive and thrive, it increases, at least a little bit, confidence in the Web of Relationships worldview.

With more confidence, even more experiments and projects will happen, leading to still more success and even greater confidence. I'm sure you are familiar with some of the projects driving this feedback loop. Perhaps you are even involved in some yourself. Here are a few examples that occur to me:

- Worker-owned cooperatives and other collaborative economic models
- Mutual aid efforts
- Regenerative agriculture
- Ecosystem-based adaptation
- Policies, like climate corps, that aim to improve well-being and equity along with climate protection
- Multiracial coalitions

- Truth and reconciliation efforts
- Returning land to Indigenous peoples
- Environmental health initiatives

When these projects deliver results, collective confidence in the Web of Relationships worldview grows a little, inspiring new experiments and creating the possibility for even more confidence. Remember that reinforcing feedback loops are like snowballs rolling downhill. They may start small and slow, but they pick up speed. They produce slow change at first, then a seemingly sudden upswing.

Reinforcing feedback loops can be activated and accelerated. Knowing these two loops are at play to potentially weaken dangerous worldviews and bring alternatives to greater prominence, we can become active participants in worldview change.

The Messiness of Worldview Change

Like pretty much everything else about complex systems, worldview change isn't simple, straightforward, linear, or easily controllable. I believe grappling with worldviews is an essential part of the work of creating a livable future. After all, we can't get very far with a worldview that doesn't fit the world. But just because it is important doesn't make it easy. Here are some factors that make grappling with worldviews challenging. Do these sound familiar? Would you add others to the list?

Pointing out anomalies can be scary. When a way of thinking seems so fundamental and so shared, it can be frightening to challenge it. We are social creatures, and breaking with family or community about core beliefs can have consequences. Many of us have experiences of holiday meals made tense by differing views about race or climate change around the dinner table or friendships floundering in a clash of worldviews.

Advocating for a different worldview can make you feel vulnerable too. As a young sustainability professional early in my career at a gathering of well-known thought leaders, I said that I didn't think capitalism was making people happy. "If we put forth a different vision, I think we could really get traction," I said sincerely if a little idealistically. I still remember when the most famous person in the room started to chuckle. Soon a handful of others joined in before the conversation shifted uncomfortably on to other topics. I had poked at an unquestioned assumption among the group and experienced some not-so-subtle pushback.

Almost two decades later, I still remember how my face flushed, how I moved to the back of the room at the first chance I found and didn't speak much for the rest of that daylong event.

I still stand by my words that day, but the memory still makes me cringe too.

Sometimes it is tempting to skirt the issue and steer away from worldview terrain to "keep the peace" or avoid pushback. But I think it is worth considering another possibility. You never know where your words might land. As the cracks in the façade of the idea that the world is merely a collection of objects grow bigger, the odds increase that others, even others who might surprise you, have their own doubts.

Worldviews can be actively propped up, sometimes with massive amounts of wealth and power. Those who benefit the most from the systemic structures derived from the Collection of Objects worldview have a vested interest in keeping that worldview supported and unexamined. Money and power are devoted to campaigns like climate denial, obfuscating key anomalies stemming from the Collection of Objects worldview. Media, advertising, and entertainment offer up glamorized views of lives oriented around consumption and consumerism, glossing over the consequences of mass consumption on a global scale.

It's natural to feel small in the face of such big forces, and I don't intend to minimize them. Still, at the end of the day, anomalies will continue accumulating the longer we stay on the current course. The oceans *will* warm and rise. Plastic *will* accumulate in the oceans and microplastics in our bodies. It's worth speaking out about the anomalies we see, even if we feel like small and lonely voices, because our words can give courage and clarity to others. As the distance widens between the stories that prop up the Collection of Objects worldview and people's own experiences, I believe we can help each other make sense of the dissonance.

Worldviews tend to be self-reinforcing. There is a lot of tension wherever worldviews clash. This tension can be reduced by seeking out sources of information and authority figures who confirm our own sense of how the world works. Social media bubbles and customized news feeds can exacerbate this tendency. I don't know a special formula to communicate across these bubbles, but I think it is good to be aware of this tendency and experiment with ways to break through such bubbles, whether cultivating a wider set of friends, changing up your reading list, or cultivating curiosity about how others see the world.

Sometimes the response to anomalies can be to hold on tighter to an existing worldview. Many of the anomalies arising are scary and destabilizing. They include resource shortages, rising prices, economic uncertainty, and more frequent disasters. While a systems thinker might see these as good reasons to get to the heart of worldviews and transform systems, not everyone sees it this way. It is probably more typical to respond to anomalies from within the suppositions of the Collection of Objects worldview. If there might be shortages on the way, be sure you can get enough for yourself and those closest to you. If there might be climate migrants on the move, build protective walls to keep them out. And so on.

To me, this potential for responding to destabilization out of the Collection of Objects worldview is one of the most important and urgent reasons to talk about worldviews in the first place. Worldviews are powerful no matter what. But they are more powerful, and potentially more destructive, when they operate in the shadows and when we don't expect their dynamics.

Worldviews can't speak for themselves. They need us to embody them in the actions we take and how we explain those actions. We can embody and articulate the Web of Relationships worldview. Whether it's composting food scraps in your backyard, designing a pilot project to restore hundreds of miles of coastline, or setting up mutual aid networks during a crisis, can you link what you are doing to worldview? "We do it this way," you can say, "because everything is connected, and neglect of those connections has put so much at risk."

The beautiful thing about change driven by reinforcing feedback is that you don't have to do it all. There are dynamics already at play, spinning along. You can add your energy and your intentionality to that spinning. You can help accelerate the process.

Whether you're a teacher, parent, boss, writer, or a movement member, you will notice anomalies and successes everywhere. Once you start to work with them, you can be a participant in processes with the potential to elevate a worldview that better fits the world and hasten the departure of one that does not fit so well.

Steering Systems with Multisolving

The idea of steering systems and the idea of multisolving go hand in hand, and the potential for both is inherent in the structure of complex systems.

By steering systems, we can influence key variables to shift in healthier, more resilient, and more equitable directions. If such shifts deliver a solution to only one problem, of course that would not be multisolving. Those shifts can also solve one problem while worsening the situation elsewhere in the system. Clearly that's not multisolving either, though it is sadly common. But if you steer systems with the goal of solving multiple problems—maybe some short-term and some long-term, or maybe some close to home and some at planetary scale—then interesting possibilities emerge. Then you aren't merely steering, you're steering toward multiple goals at the same time. With intersecting crises hitting harder and harder, this is a sweet spot we need to get better and better at inhabiting.

What does it take to work together to multisolve? What are the ways of working that enable steering toward multiple goals at once? Those are some of the questions we will take up in the next chapter.

? Questions for Reflection

- Set aside some time to envision wild success in some area of your life or work. What does it look like? Be specific. What do you really want to see? Capture your vision in words or art. If you like, share it with someone else to give it added power and specificity.
- Jot down some of the simple rules you notice for how your family, company, or team operates. Which ones do you like and want to continue to foster? Which ones would you like to let go of or change?
- Do you agree that systems can be steered even if they can't be controlled? Have you ever steered a system? What happened?
- In your world there are likely some systems that operate according to the Collection of Objects worldview and some that operate according to the Web of Relationships worldview. Some might even be a mixture. See if you can list a few examples of each worldview in your daily life.
- Pick a crisis or problem that you are dedicated to addressing. Are there ways the Collection of Objects worldview might contribute to that crisis?
- Think of an innovation that is bringing health, healing, or hope to you or others. Does it align with a Web of Relationships worldview? Why or why not?

Ribbons

For the ribbon cutting,
they chose high noon on
   a bright day,
the sun flashing off the scissors
   in the mayor's hand.
Grins for the cameras,
   tours on the hour.
"One o'clock for the rooftop gardens!
   Lemonade at the playground!"
All that is missing is a
   brass band you say.
But you are laughing, and I am too.
We see them all, the
   invisible ribbons,
stretching back, stretching between.
One is pale pink,
leading forward from that day
   in the basement of city hall
when Al threw up his hands
   and knocked the tray of
   donuts to the floor.
It was an accident, but
   also, frustration.
Then he told us about his little
   sister who almost died.
It was an asthma attack.
His voice broke and Vanessa's
   heart cracked open.
We could all see it,
a soft green ribbon.
Vanessa's boss made that grant,
a life-giving ribbon,

like a vine or a blood vessel.
Between you with your
   binders and protocols,
and me with my artist's
   messy hands
there is a ribbon, tight and
   strong, robin's egg blue.
Between Iona "how many jobs can
   we guarantee" Williams,
and that guy Steve who
   came only once,
but grabbed the purple marker
   and scribbled the answer
   we'd been missing
onto a crumpled piece of
   flip chart paper,
and Miss Marley from the
   neighborhood who
   scared us all at first
but now brings us lemon
   squares on Tuesdays
there are ribbons.
They are colorful, frayed,
   clean, crusty with donut
   icing, everything.
Thick as rope, thin as sewing
   thread or babies' hair.
The mayor can cut as many
   ribbons as he likes.
I prefer braiding them,
weaving until we have something
   so strong it will hold us all.

CHAPTER NINE

Multisolving in Action

The potential for multisolving may be inherent within systems, but it doesn't just happen on its own. Someone—a single individual, a group, or a coalition of organizations—needs to steer toward multiple goals.

This process of steering requires several activities. Some of these may happen concurrently, some in succession, and some in recurring cycles. Some of these activities are more self-evident than others, but they are all important. Figure 9-1 groups these activities into three categories likened to parts of a dandelion plant: those that take place below the surface, those carried out above the surface, and those that ripple outward.

The most obvious parts of a dandelion grow above the ground: the leaves and the bright yellow flower heads. The easiest parts of multisolving to see are like the flowers of the dandelion—they happen within plain sight and are easy to recognize. Trees can be walked under in a park. Bike lanes can be cycled on. You can see the wetland, restored, or the preschool fully operational.

More often overlooked, but still above the surface, are actions that are analogous to the dandelion's leaves and stems. These are the interventions in the system—the investments, campaigns, programs, and policies—that yield visible change in the system, the flowers of the plant. The leaves and stems are like the committee meetings, the listening session, the policy lobbying. They do the work that allows the flowers to bloom.

But the leaves and flowers aren't all there is to a dandelion, and taking action and getting immediate results aren't all there is to multisolving. There are ripples of change, represented by the top layer of the diagram, where the dandelion seeds are being blown far and wide by the wind. One multisolving project I was involved in had many discussions among its members about racial justice. One participant, impacted by these discussions, went back to her mostly white-led organization and invested in a racial equity training for her board and staff. Their projects then began to include more of a focus on equity. That's an example of a ripple of change.

To multisolve in the first place, the capacity to act together must be nurtured. People need to come to trust one another and to develop shared

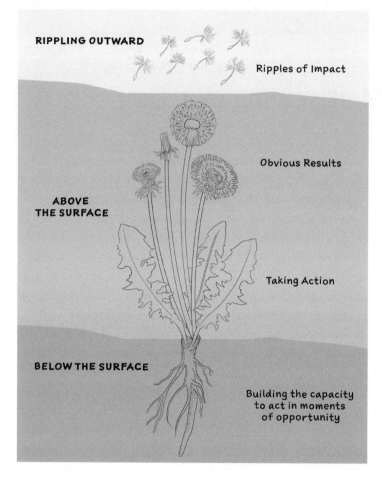

RIPPLING OUTWARD

Ripples of Impact

Obvious Results

ABOVE
THE SURFACE

Taking Action

BELOW THE SURFACE

Building the capacity
to act in moments
of opportunity

Figure 9-1. A dandelion plant offers a metaphor for multisolving. Below the surface, less-visible work is done to build the capacity to act together. Above the surface, people take action, make investments, and enact policy that produces visible results. Other results ripple out over time in chains of cause and effect, but these can be harder to notice than the visible results because they are more spread out in time and space.   *Illustration: Molly Schafer.*

strategies and visions. This "below the surface" work is represented by the diagram's root system. While vital, it is missed by many people. In fact, I've noticed that multisolvers themselves don't always consider this work as "action." And if practitioners miss it, it is even harder for funders and other supporters to recognize these activities. The below the surface layer includes things like convening people, growing trusting relationships, fundraising, learning together, and aligning vision and values. It's often unglamourous, but the "above the surface" work would be less achievable and less effective without the nurturing going on in the "soil."

In the sections that follow we will look first at the most visible part of multisolving—taking action and generating visible results. Then we will look at ripples of change, and finally the patient tending "below the surface."

Taking Action

Above the surface of a dandelion are the leaves and stems. They do part of the "work" that is needed for the dandelion to produce results (blossoms and seeds). Think of this work as the visible actions and interventions in systems being carried out by multisolvers.

In a general book on multisolving it is challenging to talk specifically about "taking action" because multisolving is dependent on initial conditions and sensitive to local needs and opportunities. It is implemented very differently in different places. Still, in the midst of all of this diversity there are some common themes.

Often a cross-sectoral network is the "unit" that is acting. The exact form of the network varies from place to place. Action may be arising from organized communities or from business partnerships. It might be coming from a coalition of government officials, citizens, and nonprofits. Those taking action may be volunteers or paid staff or both. The funding may come from foundations, individual donations, corporate grants, ratepayers, customers, government programs, or some combination of these.

Step back from the specific forms and what do you see? People spanning silos. We noticed this in creating a series of ten profiles of projects working at the intersection of health and climate change.[1] One profile was of Operation TLC, an energy efficiency program in a group of UK hospitals. That project involved a university partner, hospital administrators and staff, the Department of Health, and a nonprofit focused on energy efficiency. We also profiled ProAire, a project in Mexico focused on improving air quality. That project included even more partners: five academic institutions, seven businesses, dozens of state, local, and federal government agencies, a health care organization, and at least ten civil society organizations.

The actions fit their context. What are the needs in a particular place? What are the capacities? What are the resources? The geography? The culture? Who is ready to take action? What is necessary and useful in any situation will differ depending on the system's history.

Borrego Springs, California, was suffering power outages due to wildfires and storms that cut off the transmission lines from the distant power plant that generated its electricity.[2] The loss of power was dangerous because temperatures climb high in the region. The community had a high percent-age of older residents who needed air-conditioning to stay cool and refriger-ation to protect vital medications. Partners from the University of California, the local chamber of commerce, a local conservation organization, several energy companies, the US Department of Energy, and several other gov-ernment agencies worked together to install a solar microgrid, lowering greenhouse gas emissions and making the community more resilient. That particular solution fit the specific place, local conditions, the needs of local people, and the changing circumstances.

Multisolving actions change systems. Because multisolving is so specific to particular times and places, it is not very helpful to list all the possible actions that could multisolve. What's more helpful is to say that wise multisolving actions employ the principles of systems change we have discussed in the opening chapters of this book.

A multisolving project might reduce food waste. That's reducing an outflow of a stock of usable food. It's also reducing an inflow of CO_2 into the atmosphere since agriculture is a major source of greenhouse gases. Returning land to Indigenous stewardship is an action. It could reduce the outflow of biodiversity loss and increase the inflow of sequestered carbon. It's also a way to contribute to a worldview shift from one of domination and supremacy to one of relationship and interdependence. When it comes time to strategize about the particular multisolving action best suited to you and your partners, my advice is to try to see the system and plan your interventions with the stocks, flows, and feedback loops you identify in mind.

Often, multisolving action happens during "moments of opportunity." A new source of funding is announced. A change in company leadership puts new possibilities on the table. A planning cycle opens up. Public opinion shifts in response to a disaster.

Moments of opportunity present a paradox. Often, they are fleeting. Yet capacity to act within them requires time to cultivate. If you wait to start the process until a moment of opportunity emerges, you may miss the window. Looked at superficially, multisolving actions can seem almost like magic, like the stars aligned and big progress happened. More often, the groundwork for that moment of action required months or years of meeting, planning, and trust-building among partners.

Like moments of opportunity, moments of danger are also times for multisolving action. During the COVID-19 pandemic, many multisolving projects pivoted to help address the immediate needs of their communities. One small grants project we helped organize awarded grants just before the pandemic began. As the region entered "lockdown" we saw several of the small cross-sectoral teams shift their plans in response. One project took funding meant for park beautification and used it instead to teach senior citizens how to use their phones to access telehealth appointments. Watching these pivots taught me that connections that are built for one set of goals can be repurposed when a new need emerges.

Windows of opportunity can sometimes be pushed open. When steady power building and organizing leads to an important message being heard or newfound access to political power, then the window of opportunity isn't opening just by chance or due to factors outside of a project's or network's control. Instead, through strategy and hard work the window is intentionally opened up.

In either case, whether the window is a surprise or something long worked toward, there is an interesting interplay that is important to pay attention to. On the one hand, moments of opportunity (or danger) are fleeting and often unpredictable. On the other hand, the capacity to act in those moments is not assured. Building the capacity to act in a moment of opportunity can be a slow process that can't be rushed. A lot of that building happens below the surface, and I'll have more to say about that later in the chapter.

Before we do that, let's look at the results that can emerge when multisolvers take action in moments of opportunity.

Obvious Results

Wherever multisolvers take action, and in whatever configuration they organize themselves to act, those actions produce results. Most multisolving projects will, in my experience, fit into a few broad categories that I think of as "solution clusters." The results each cluster produces, like the attention-getting bright yellow flowers of the dandelion, are often the most visible part of multisolving.

Restoring and protecting ecosystems. When approached with a multisolving perspective, restoring and protecting ecosystems can safeguard biodiversity, improve air and water quality, build climate resilience, boost livelihoods, protect cultural traditions and knowledge, sequester carbon, provide meaningful work, and enhance food security. Examples include reforestation, water conservation and retention to replenish groundwater, regenerative agriculture that protects and restores soil fertility and microorganisms, and land conservation, especially when it is led by local communities, including Indigenous peoples.

One type of multisolving that I love from this category is the living shoreline, like the Goldbug Island project off the coast of South Carolina, where local businesses, the Nature Conservancy and other conservation groups, and more than two hundred local volunteers worked together to restore oyster habitat offshore.[3] The oysters help filter and purify water and support more than one hundred other fish and invertebrate species. The restored vegetation helps sequester carbon and protects the area from sea level rise.

Ecological, human-scale community design. Designing the built environment with multisolving in mind produces gains in mental and physical health, reduces energy use and climate pollution, boosts local economies, protects air and water quality, and promotes equity. Examples include walking and cycling infrastructure; green infrastructure to prevent flooding, cool cities, and support people and biodiversity; public transportation; community land trusts; and energy-efficient, dense, affordable housing.

Just Communities is a groundbreaking project that promotes equitable and sustainable urban development, and I had chance to watch the project develop as a member of its advisory board.[4] Its formation united two areas of expertise when one organization acquired another. The acquiring organization was Partnership for Southern Equity, which has expertise in improving racial equity in the American South and across the country. It acquired Eco-Districts, a nonprofit that certified green-building experts who focus not just on individual buildings but on creating whole green and sustainable neighborhoods. After the merger, the teams from both organizations worked for more than a year to create a new training and certification program that united green design, racial equity, and community leadership into a cohesive whole. Starting in 2024 communities across the country can follow the principles outlined in the Just Communities Manifesto to guide their growth and development, finding synergies between racial equity and ecological design, and they can identify practitioners who have been trained in the approach to help them.[5]

Clean energy infrastructure. Creating highly efficient clean energy infrastructure allows people to meet their needs without creating climate pollution and provides good jobs, climate resilience, and reduced energy burden (the fraction of income someone spends on utility costs). Examples include energy efficient buildings, water efficiency, wind and solar energy infrastructure and the transmission and storage upgrades that support them, and microgrids to increase access to electricity. Nashville's Home Uplift program offers a good example of this sort of multisolving.[6] In this project the mayor's office, AmeriCorps volunteers, local businesses, and an electric utility collaborated to weatherize homes. The project focused on the Nashville Promise Zone, the area with the highest poverty rates in the city. In addition to reducing climate pollution, the upgraded residences showed an average 25% reduction in their energy bills and reported that their homes were more comfortable and livable.

Human- and Earth-centered economics. Shifting economic structures and incentives with multisolving in mind can improve equity, reduce poverty, protect resources, increase well-being, and increase climate resilience. From sharing tools, materials, and resources to reusing and repurposing to make the most of materials, to shorter working weeks and telecommuting, there are lots of ways to adjust economic models to create wins for people and for nature.

Ensuring civil, human, and other rights. Protecting universal human rights is a multisolving solution that improves the equity and overall health and well-being of societies. It also is a potent protector against environmental destruction and unsustainability because every river, forest, neighborhood, and stretch of coastline is precious to its community. If every community had the power to protect itself, we'd have a world with fewer sacrifice zones and more vibrant ecosystems and human communities. Protecting gender rights, promoting the education of girls and women, and protecting democracy and voting rights are all multisolving solutions that improve many aspects of a system at the same time.

Conflict resolution and peace promotion. Money spent on war and conflict could be redirected to life-sustaining activities, from health care to education to ecological restoration, and the devastating ecological impacts of war could be avoided. Ideas like departments of peace (instead of "defense") and education that fosters peace building and conflict resolution would spawn multiple benefits. Countries are experimenting with ideas like climate corps that provide young people with service opportunities and communities with support in climate change adaptation. Around the US, activists call for reallocating budgets for policing toward investments in education, community well-being, mental health, and conflict resolution.[7] Costa Rica stands out as an example at the national level. Having made the decision to not fund an army, the country reaps a "peace dividend" by being able to devote more resources to goals like education and development.[8]

Rippling Outward

Not all changes that emerge from multisolving are immediate or direct, as represented by the dandelion seeds in the "rippling outward" layer of figure 9-1. Changes created in a multisolving project can lead to further changes and enable new possibilities and innovations, even distant ones. The timescale of these changes can also be much longer than that of the initial intervention. A restored ecosystem might result in some species bouncing back quickly while others take decades. That restored ecosystem might then enable jobs, and the jobs might reduce food insecurity, and so on. Not every multisolving project will be able to identify or foresee all these ripples of change, but they are worth looking for.

Sometimes the ripples that have been created are revealed by later events. Lakshmi Charli-Joseph is a researcher in Mexico we interviewed for a research paper on multisolving.[9] She and her colleagues brought together farmers, residents of informal settlements, artists, and others to explore collective solutions for an endangered wetland. Again and again, they gathered together to listen to one another, even hiking together to the top of a dormant volcano to look down on the wetland together. After the formal close of the project their region was struck by a strong earthquake. In the months that followed, Charli-Joseph and her colleagues observed that relationships between participants in the wetland project were helping each other navigate the recovery process. A project that had been focused on wetlands had created a ripple of strengthened relationships that made disaster recovery a little bit easier.

It is probably the case that you will never know about some of the changes your multisolving contributed to. But that doesn't mean you shouldn't look for evidence of ripples from your work. Because of how challenging it can be to document the ripple effects of multisolving projects, I suspect that multisolvers consistently underestimate their impacts. But one thing is sure: if you don't look for ripples, you won't find them!

Below the Surface

Being able to act in moments of opportunity requires patient, below the surface work.

Take the Just Growth Circle in Atlanta, a project I collaborated on in its early stages with Partnership for Southern Equity.[10] That project brought together advocates for green infrastructure, racial justice, health, climate action, and housing in an ongoing series of gatherings. The group painstakingly drafted a core set of consensus values. Members drew maps of systems and interconnections and presented "project clinics" about their work at the intersection of equity, water, climate, jobs, and conservation. Each meeting had a "needs and offers" section, where people could ask for and offer help. All of this activity was weaving a web of interconnection: shared ideas, shared visions, and trusting relationships. Even so, from time to time participants and funders would have trouble seeing what the group was "accomplishing." This didn't look like the sort of specific, coordinated, top-down plan people are more familiar with. It was what I'd call below the surface work, the kind that enables all sorts of more visible action.

What emerged from it? Well, the story is still unfolding, so I can only tell you part of it, but here are some things we observed that were helped along by the activities of the Just Growth Circle.

Protecting against displacement. In a watershed planning process, Circle members secured commitments to protect against displacement for people in marginalized neighborhoods.

Influencing spending. Circle members were involved in updating the Watershed Department's billion-dollar capital improvement plan, making sure that equitable hiring and procurement provisions were part of the plan.

Influencing infrastructure plans. When the city released a green infrastructure plan, the core values of the circle—including the principles that investments should protect marginalized groups and impacted communities should have a say in infrastructure decision-making—were included.

Leadership development. Over the years members of the Circle have collaborated on leadership development efforts, including leadership programs for residents and youth and a Development Watch Academy that equips residents to monitor and influence development activities in their neighborhoods, part of a strategy to combat gentrification and displacement of long-term residents.

Accessing large government grants. The connections forged within the Just Growth Circle laid the groundwork for a successful application for a multimillion-dollar US Forest Service Grant for a large-scale urban forestry project.[11]

The successes of projects like the Just Growth Circle made me very curious about what was going on beneath the surface to develop the capacity to multisolve. Here are a few of the activities I've come to believe are most important.

Sensing the System

Multisolving projects are path dependent. So even if their starting points are similar, no two multisolving projects will develop in the same way. Chance events and the unpredictability of emergence will see to that!

You'll need to become a learner and an observer of the systems you hope to steer toward multiple goals. Part of being able to act in moments of opportunity is developing a sense of dynamics, patterns, and momentum in the systems you are working within. Here are a couple of questions to ask to help develop that sense.

What opportunities are emerging? Complex systems always feature motion and change. Rarely can you predict exactly when specific windows for action will open, but you can be sure such opportunities will arise, whether it's a call for proposals, a funder asking for new ideas, a vote on a key ordinance at city hall, a new business ready to make a big investment and looking for partners, or your boss adding a new objective to your to-do list.

If you are already part of a group of multisolvers, scanning for emerging opportunities is a good habit. Just ask yourselves what opportunities are emerging from time to time. What you have built together—the connections, the coherence around values, the learning from past action and experience—is just what you need to respond to emerging opportunities. A connected system, like the one you have been cultivating, is able to sense and respond. You can shift attention, design the plan, write the grant, mobilize for the hearing. But before you can do any of that you have to sense the opportunity or the need.

People working in different parts of the system will naturally be aware of different needs and opportunities. You can tap into this resource if you make space for regular check-ins where people can discuss emerging opportunities. That space could be a few minutes at the beginning of every meeting to ponder an open question like "what opportunities are you sensing?" It could be a hashtag on social media, a section of a newsletter, a topic on an email mailing list.

Think back to our discussion of balancing feedback loops and how they can activate gap-sensing mechanisms that change the state of a system. An opportunity is a potential gap-closing mechanism. But a feedback loop is only as strong as its weakest link, so an unnoticed moment of opportunity means the loss of a potential balancing feedback loop.

Because moments of opportunity are fleeting, the time to "spin up" a balancing feedback loop is limited. Scanning, naming, and sharing opportunities as they emerge is one way to speed up a system's response time.

What needs are emerging? After Tropical Storm Irene hit my home state of Vermont in 2011, there was an urgent need to replace flooded mobile homes. That need was met with multisolving. In addition to replacing damaged housing, the project partners behind the "Irene Cottages" also created a design for energy efficient homes that retained many of the architectural elements of the damaged housing.[12] In addition, they designed a financing mechanism to help lower-income Vermonters afford the slightly higher upfront costs by balancing those costs with future energy bill savings. That's multisolving for climate resilience (relocating from the floodplain), climate protection (reducing emissions from the buildings sector), and economic equity (lowering barriers to accessing energy efficiency benefits). And it started with noticing multiple needs.

Whenever you notice a need, before plunging into single-focus action, take some time to reflect: are there other problems or crises that could be addressed at the same time?

Multisolving bundles together problems that are often addressed in isolation. So if you notice one need, take the time to look around and see what other ones might be connected to it.

Given the dynamic nature of complex systems, especially in the times we're living through, many needs are constantly emerging. Scan for them. What's needed? What's emerging? Where is help, care, assistance needed? Budget cuts, election results, climate impacts, economic shocks can all be sources of the kinds of needs that could be met with a multisolving response.

New needs might also show up on your desk as part of your job. They might appear on the agenda of your volunteer organization or as a problem faced by a market your business serves.

If you have been multisolving with partners for a while, new needs will also come to your attention. Even while you have active programs in place addressing needs you've identified in the past, stay alert for shifting or emerging needs because they can signal either new opportunities for multisolving or new potential benefits that can be bundled into your ongoing work.

A crisis can be the precipitating event that leads to a gathering of people across professional, cultural, and intellectual silos, combining them into a force that can address more problems than the one that led to the gathering in the first place. Of course, most crises are not met with multisolving. But greeting a crisis with a question—what else might improve by virtue of how we respond—can open the door to multisolving, and that crisis response can put systems on a better footing.

That was true in Vermont's Irene response regarding more than floodplain housing. In 2023, twelve years after Irene, Vermont was hit by heavy summer flooding, reminiscent of the tropical storm. Once the immediate emergency was over, the recurrence provided for an interesting period of reflection. How did the state's infrastructure fare? Were there signs of multisolving back in 2011 that had the state better prepared for a disaster in 2023?

In 2011 Sue Minter was Vermont's Tropical Storm Irene recovery officer, and in a 2023 interview she was able to point to some encouraging signs. For instance, in 2011 when six hundred miles of roads—including all of the state's major east–west roadways—were flooded out, the state's Transportation Department's Emergency Operations Center was flooded out too, hindering the response. In 2023, Minter said, "That's very different from today, where I think the state is more prepared, working together in an interagency system."[13] That's a nod to multisolving—forging connections across agencies to better coordinate action.

The state multisolved in another way: Minter and her team were able to connect the dots between emergency response and long-term climate adaptation, with the result that infrastructure fared better in 2023 than it did in 2011. In that same interview Minter said, "We designed our recovery with resilience. For example, we've worked so hard on the state system, and also the town's system to ensure that we didn't just replace culverts and bridges the way they had been before, but we built them at a larger span for bridges, larger volume for culverts. I think that has had a tremendous impact on the system where, after Irene we lost 200 bridges overnight. But as far as I've learned, so far, only two bridges along the state highway system have been impacted."[14]

In an era of destabilization, crises are one thing we can be sure will be coming. Some of them may best be met not with a narrow response but with a wider gathering of allies and a long-term view.

Deepening in Vision, Values, and Simple Rules

We've discussed how vision, values, and simple rules offer ways to steer complex systems. In your multisolving projects you will be operating in conditions of uncertainty and flux. New partners may join, or be interested in joining, your efforts. Old partners may move on. You'll have successes and make mistakes. Through it all, if you are clear about and aligned with your values, vision, and simple rules, you will have a greater sense of stability and increased ability to steer.

When moments of opportunity appear, you will be able to evaluate them relative to your deepest visions, values, and simple rules. But only if you already understand them. Especially in a crisis, you may not have time to do the patient work of grappling with guiding values and principles. That's why, below the surface, one key to building the capacity to multisolve is to make space for that patient work from the start. You can take the time to explore and define visions, values, and simple rules early and update them as the makeup of your project partners shifts, as the wider system changes, and as you learn and grow.

There are many ways to engage with values and vision. You can conduct formal vision sessions and share individual visions in words or art, exploring their points of intersection and divergence. You can use a consensus process to determine what values you have in common and what language works best to express them.

I observed one powerful way to work with values in the Just Growth Circle, an innovation suggested by our partner Tina Anderson Smith. Members of the Circle were eager to share their work with one another, and we devoted a part of each meeting to "project clinics" where one member would share their lessons and struggles in a current project. Since the group had already developed a consensus values statement, Tina suggested that the focus of the project clinic be on two questions: How do the shared values of the group come to life in the project being shared? What are the struggles or challenges in embodying those values?

These questions transformed the project clinics from simple "report-outs" to opportunities to grapple with what it means to put values into practice. It opened interesting questions, like what to do if you overhear a racially insensitive comment in a meeting. Or how to navigate the tension between a tight timeline and commitment to community input, which takes time. If one participant was having a particular struggle, others might have encountered it too or even found a solution.

Simple rules can help codify the culture of your project as well. Some examples of simple rules that could help a multisolving project thrive include:
- We share the work and the resources.
- We are transparent with finances.
- We respect lived experience as much as formal training.
- We make space for challenging conversations.

The way you design meetings and make decisions as a group will provide dozens of opportunities to embody simple rules that are important to you.

Practice in the day to day will prepare you so that, in unpredictable moments of influence, your efforts will be skilled and coherent.

How do we share funding? Who will give the next presentation? How much effort should we put into responding to the issues raised by young people? Questions like these offer opportunities to increase the coherence of your network.

Visioning can also help ease the way into multisolving. Start by imagining that an action you've been contemplating within a system has already happened and was a tremendous success. Picture the new policy fully implemented, the technology deployed, the forest protected. Now ask yourself what else has improved. Maybe the air or water is cleaner. Small farmers have higher incomes. Soil has been protected. People have access to better-quality food or more affordable housing. Write down as many improvements as you can imagine. Don't be surprised if some (or even many) of them are outside of your area of expertise.

Now you have the start of a simple map for multisolving. Each possible improvement represents a constituency who could be just as enthusiastic as you are about the action. Aim to include, eventually, at least one "champion" of each of those benefits. If you're looking to create a farmers market, your list of potential partners may include farmers, community garden stewards, the local food pantry, churches, public nutrition experts, climate advocates, consumers, and educators. That's a good starting list of folks to invite for coffee or to an exploratory conversation.

It's okay if you don't know people connected to some of the potential improvements. That just means you have another question to ask: Who is doing the best work around here in nutrition (or hunger or regional climate policy)?

Nurturing Connections

Connections—between people, between organizations, between ideas— are part of what makes acting in moments of opportunity possible. In her book *Emergent Strategy*, adrienne maree brown points out that change happens at the speed of trust.[15] Creating partnerships takes time. So does learning about the needs and constraints of people acting in other parts of systems.

Of course, connections are built when people act together. You can probably recall several moments in your own life and work when coming together to do something challenging built lasting friendships or alliances.

Like so much of multisolving, the building of connections is organic and path dependent, so there's no one "right way." That said, a few questions can help focus attention on what connections might be helpful to attend to.

One question is, where are we limited? Multisolving projects involve a lot of learning by doing. In the process of documenting your successes and positive impacts, you may also discover places where you didn't have your desired impact. You may discover areas where neither you nor your partners have the required expertise. Your limitations can point you toward a place where you need to nurture additional connections.

The Just Growth Circle started as a network that included many leaders focused on land conservation, green space, water, and health. After some work together, the initial group identified "green gentrification" as a concern. Green gentrification is the process by which adding sustainable infrastructure to a community drives up property values and leads to the displacement of longtime residents.[16] None of the participants wanted to be responsible for such a harmful side effect, but they began to realize that their work could lead to this outcome. They recognized their need to know more. They wanted to be able to be effective in new-to-them sectors such as affordable housing. With this in mind, they expanded the group, inviting experts on affordable housing, community development, and gentrification to join the project.

When you start your multisolving effort you may not know all the parts of the system that will matter, but that's not a reason to not start. If you listen to it, the system will inform you of what (and whom) you need to include.

Noticing where you are running into limitations can be a step toward expanding the effectiveness of your multisolving. Maybe you don't have sufficient voter support to influence a policy target. Or you don't have enough funds to disperse small grants even though you have strong evidence they could have a big impact. Each of these limitations might be a signal to expand your circle. At the very least, each limitation is telling you about a dimension of multisolving that you haven't yet incorporated fully into your work.

Looking out for who seems frustrated is another way to notice connections ripe for the making. Every person with a dream that is blocked by silos or jurisdictions is a potential multisolving partner. Maybe you yourself are frustrated at missed opportunities. Why not try reaching out to others to see what might be possible in partnership?

With so many of us working within silos, these frustrations sit close to the surface. Everything that a silo prevents someone from doing is something that might be possible with a partner on the other side of the division.

Many of us hold values that are in some amount of conflict with our official roles in systems. Say racial equity is a strong personal value of members of your team. But they are trained as naturalists. Your organization is known for expertise in biology, not social justice. That internal tension is the perfect push to design multisolving into your next conservation project. What new skills would your organization need to develop and who would you need to partner with to implement antiracist conservation projects?

Another powerful question for nurturing multisolving connections is, who is impacted? Multisolving works when it includes the people who will live with the results of the work. Maybe you are one of those people, or you already partner with them. If not, asking who is impacted is a key part of building the connections needed to multisolve effectively.

If your project is a park, have you reached out to the neighborhood association? How about the businesses adjoining the proposed cycling path? Who will live with the impacts of your microgrid, new agricultural practice, or more energy efficient homes?

Sadly, in traditional approaches it is common for change to happen *to* people and communities rather than be shaped *by* them. Multisolving includes those most impacted in the core of project design and decision-making from the beginning.

"Who is curious?" is another question that can help you notice opportunities for growing connections that will strengthen your multisolving. Multisolving happens at the intersections between disciplines, sectors, and jurisdictions. To see and act on opportunities, reach out to people who excel at flexible, curiosity-driven thinking, people who won't mind being students again, and people who are happy to explain their part of the system in a way that nonexperts can understand.

How do you find such people? Some of them you likely already know. Think of colleagues you've collaborated with in the past. Who was easy to work with? Who was always interested? Who was energized by learning and trying on new perspectives?

Others you will find by referral. Don't just ask for the expert on food security in this province. Ask for the expert in food security who is also curious, humble, and excited by learning. People will know.

Finally, as you meet people who are drawn to your multisolving project, don't hesitate to interview them a little bit. Steer clear of people who seem sure of all the answers, and be on the lookout for folks with lots of questions!

You can create the conditions for curiosity, exploration, and a spirit of learning in building your project's culture. But it helps to have a core of participants who are inclined in that direction in the first place too.

So far, the questions in this section have been about noticing where fruitful connections might be made.

Equally important is cultivating conditions where new connections can form and where existing ones can strengthen.

Almost everything you do in a multisolving project is an opportunity to strengthen relationships. The first invitation to a conversation. How you make project decisions. The mechanism for sharing resources. Sharing information and opportunities. With every decision you make, whether it's creating the next agenda or designing a pilot project, "how can we support connections forming" is an important question to ask.

In fact, what you do may often be less important than how you do it. Remember, you are trying to create conditions for emergence. Emergence happens when new connections are made in a system. Introductions around a circle are a chance to make connections. Allowing people to share their frustrations, values, and uncertainties is another way. Working side by side forges connections. So does sharing ideas.

Time spent together face to face helps, such as projects or field trips or other opportunities to connect and learn from one another. Establishing ways for resources to flow from member to member within the network also helps build connections. Resources may be money, they may be tools, and they may be knowledge or access to networks.

One metric you can track concerns the direction of flows. Are resources flowing in all directions? Are large well-funded entities able to share with smaller organizations that might not have the same access to resources? Are community-based participants, or those without formal credentials, respected as leaders? Are they given opportunities to be teachers?

Think also about what conditions disrupt connection. When people aren't listened to. When credit isn't given. When people don't follow through on their obligations. If requests for confidentiality aren't respected.

All along in a multisolving project, take time to explore within your network the conditions people feel foster connection. What does everyone need to feel connected to the shared work? Listen to those answers, reflect them back, and incorporate them into the way your project works.

Don't worry too much about getting the perfect representation of the whole system at first. You can grow your project beyond your initial cohort.

Multisolving is an organic process. You will learn more as your efforts progress, and it may feel natural to invite additional participants. Maybe you have the resources to convene a large group once a month for a year. Maybe you have no resources but some interesting cross-sectoral partners with whom you love to work. Either can be the start of a multisolving approach. Multisolving, and the building of connections needed to do it, is more of a way of going about your work than a particular set of steps.

Aiming for Equity

Successful multisolving efforts pay attention to equity. This attention helps prepare networks of multisolvers for acting effectively in moments of opportunity. I'll have much more to say about equity and multisolving in the following chapter, but for now, I'll share a few key points.

In the previous section we discussed how multisolving efforts are constantly nurturing connections. Inequity is anathema to trusting relationships. Inequities are a symptom of a worldview that treats people and organizations as collections of objects rather than webs of relationships. Therefore, nurturing the relationships needed for multisolving requires vigilance about equity. Equity will manifest in decisions about resources. It will show up in whose ideas are listened to and whose are not.

We've also discussed how sensing the system is a key activity that happens below the surface in multisolving. Seeing the whole system is not possible without the knowledge and lived experience of people across the system. Inequity puts such sensing at risk. If some perspectives aren't included because a group of people isn't included or listened to, the sensing of the system will be only partial. And if the capacity for sensing is impaired, so is the capacity for acting together in moments of opportunity.

Seeing and Tracking Change

Being able to see and track change is important for multisolving for two reasons. Identifying co-benefits that arise is a way to attract new energy and resources to a multisolving project. When the system changes in line with your goals, documenting the good you are doing helps bring new partners and funding to your work. That's important!

And when the system changes in ways that you didn't expect or ways that are counter to your goals, noticing those changes early is helpful for course correction, learning, and improvement.

But seeing change in complex systems isn't always easy. Often the kind of change that comes with multisolving is emergent and unpredictable, making it hard to know where to look for signs of change. When many co-benefits arise at once, it can be hard to track them all. And the changes you contribute to don't happen all at once or all at the same time. Some will be apparent quickly, but others may play out over much longer spans of time. When project teams turn over, grants come to an end, or other emergencies crop up, this slower-to-occur change can be hard to keep track of.

Change that your work contributes to can also set off cascades of change elsewhere in the system that can be easy to miss, especially if you aren't looking for them. Remember the little dandelion seeds drifting away in figure 9-1?

Tracking change involves an interplay between the "below the surface" layer of a multisolving project and the "above the surface" and "rippling outward" parts of the system. Below the surface you are doing the steady work of setting up and maintaining systems for tracking change. Many systems for change tracking can work: notes in your journals, thoughtfully timed series of interviews with participants, spreadsheets tracking metrics and measures that you have chosen in advance. Whatever system(s) you adopt, just remember to allocate time and resources to the below the surface work of documenting change above the surface, including what is happening in the harder to notice area of "outward rippling change."

Here are a few things to keep in mind as you track change emerging from your work.

Think about baselines. For your multisolving work to thrive and grow, you and your partners need to know what difference you are making. But if you don't record the condition of the system when you start out, you won't have a clear sense of your impact. That's why you need to document, as best you can, the state of the system at the start.

If you can document quantitatively, fantastic. Record as many things as you can easily measure, even things you aren't sure will be improved by the project. Not everything that matters can be measured, however. Some shifts will be in people's thinking. Some will be in how people feel about their neighborhood, their organization, their forest. Before you set out, can you record their understanding of the system, their level of confidence, their sense of themselves as leaders? Their hopes, their fears?

Consider documenting the connectivity of the system. Can you record who knows whom, who shares information or money with whom at the outset? That way it will be easier to notice as connections strengthen and multiply.

Some changes wrought by multisolving projects are so subtle and natural that people don't notice them. It can feel as though they have always known what they know or that they've always had access to the same resources and relationships they have now. Having a good baseline will help you document these changes for yourself and others.

Even if you are well into a multisolving project, baselines are still relevant. As your work progresses you may learn about new areas of the system that are affected by your work. At the outset you may not have known to measure health or happiness or fish population levels. As soon as you notice something else that is moving in response to your efforts, record its current state as best you can. Look back and document its history as well if you are able.

Remember that qualitative changes count. Not all the change you create will have a number associated with it. Interviews, photographs, and stories are also ways to capture some of the changes happening within a system.

You can also track your own impressions and intuitions. A simple notebook or journal where you reflect on meetings, gatherings, strategy sessions, or reports from the field can be a helpful resource for noticing patterns and change.

As you think about what to track, think about the multiple problems you are trying to bundle together and make sure that you have at least one method of noticing change for each of those problems.

Any documenting that you do leads directly to another part of what is going on below the surface. Documented impacts are part of what helps multisolving efforts grow. They are what attract new funders and new partners. In this way, tracking change helps you attract the resources needed for more of the "below the surface" work of multisolving and the action in moments of opportunity that springs forth from it.

A Virtuous Cycle

Often a cyclical pattern emerges in multisolving. Some work below the surface enables action in moments of opportunity. That action leads to change, at least some of which is noticed and documented. The observed changes feed back to energize important below the surface work in several ways.

By showing how the system is changing above the surface, you build confidence and access additional support for the work, enabling more intervention and more impact, which in turn attracts more support.

Observations of what is changing also feed into a better understanding of the system, increased perceptiveness about moments of opportunity and better strategies, improving impact, creating more change, and creating more learning.

During each trip around the cycle, participants can become wiser than the trip before. A cycle allows for growth (including the kind of growth in impact that is fueled by reinforcing feedback). Each trip around the cycle can be more powerful and effective than the trip before and undertaken with more numerous and more diverse partners.

And yet a single cycle that has only three stages is too simple to apply to every situation in this complex world of ours. Your cycle might branch. It might stall in one stage, or it might skip a stage. Your multisolving effort might subdivide into two projects because you are learning and following what you learn and listening to one another. You might spawn quick cycles while longer cycles continue to turn more slowly.

Though I hope the image of a cycle will help guide your efforts, listen always to what the system you are working within has to tell you. You'll find what you most need to know there.

? Questions for Reflection

- Which of the multisolving solution clusters described at the beginning of the chapter interest you most? Are there other clusters not listed in the chapter that could be?
- Can you think of a time when the types of activities that happen below the surface in multisolving networks enabled people to take effective action in a moment of opportunity? What were the most important things going on below the surface?
- Were you ever surprised to learn about a change that had been set off by your actions but took a long time to materialize or happened somewhere distant to where you acted? Write down or share this story of change rippling outward. What did you learn from it?
- Does the idea of multisolving as a cycle resonate with you? Why or why not?

Enough

Here we are,
nodes in a watery jewel.

Here we are suspended in space,
evolving around the sun.

Where we are
there's enough.

Open the dams.
They will never hold anyway,
here where everything flows.

Let everything run free.
Feel water soak into the dry spots.
Pass it in clay jugs
from hand to hand to hand
until all the thirst is quenched.

Watch fish flash upstream.

No not one child thirsty.

Revolution around the sun.

CHAPTER TEN

Multisolving and Equity

Of all the important elements of multisolving—a systems view, working with uncertainty, and tapping emergence among them—the pursuit of equity is the only one to which I have dedicated an entire chapter. That's how critical equity is to multisolving. Attention to equity is a prerequisite for multisolving, while improved equity is also a possible outcome of it. How those both are true at once will, I hope, make more sense by the time you reach the end of this chapter.

Before we get to ways to think about equity through a systems lens, some important disclosures are in order. Though everyone in a system moves through the same system, we don't each experience the system in the same way. Some feedback loops act strongly on some of us and less so on others. In the United States, for example, a Black teen and a white one might walk through the same shopping center, but the former might be not so subtly followed by the security guard and the latter might pass by without scrutiny. Some structures of the shopping center as a system are latent for the white teen and strongly activated for the Black one. Their experience of the system is different.

The fact that we experience systems differently can be a complicating factor both in our collective grappling with them and in our efforts to make them work better. It helps explain why bosses sometimes miss seeing opportunities for improving workflows and why able-bodied people might not understand the impact of an obstacle on the sidewalk to someone using a wheelchair.

All your discussions about systems will be more productive if you can remember this feature of systems. You will see/feel/experience/be impacted by aspects of the system that other people will not, and they will see/feel/experience/be impacted by aspects of the system that you will not. Sharing what you see and explicitly naming the perspective from which you are seeing can help make conversations about systems more effective and more productive.

With that in mind, I'd like to explicitly name a few of my perspectives that are likely to come into play in the rest of the chapter. I invite you to think about how my perspective has helped me see some things but how it might also keep me from seeing what you see.

I'm a middle-aged white person, educated in the sciences at prestigious universities in the United States. I have a graduate degree and have held leadership positions within various organizations for years. In these ways I'm more like the white teen moving through the shopping mall; many inequitable structures haven't impinged on me very much.

At the same time, I'm a woman in a field dominated by men. I've often been the only woman around a table or in a room, and when I entered that prestigious university, it hadn't been coeducational for long. Those experiences helped me better understand what it can feel like when inequitable systems throw obstacles in the way of some groups of people.

I've also, by virtue of studying and thinking about multisolving, had the chance to participate as groups grapple with inequity, and witnessing those hard, heartfelt, sometimes wrenching conversations has helped me better see some of the inequitable systems my own direct experience left obscured.

Finally, I'm American, and that means some systems of oppression are very familiar to me while others that are more pernicious in other parts of the world I know much less about. For those systems most relevant in your life and work, I hope that (because systems are coherent and self-similar) you will find ways to take the insights and examples offered in this chapter and apply them to your own world.

As you read this chapter, look for what I see and can articulate based on my years of moving through systems as "me" and take from that what serves, what illuminates what you might not have yet seen. But also hold it up critically against what you see that I might not. It will take all of us, from our many vantage points, to see inequitable systems clearly and to transform them into something healthier and more just.

So, what is equity when viewed through a systems lens?

I would submit that equity (or lack of it) is a property that emerges from a system's structure. The arrangement of stocks, flows, and feedback loops in a system produces whatever level of equity the system displays. The system structure influences whether people are paid equally for equal work. It shapes which groups have access to education, safe housing, and clean water and which groups do not. Regarding physical environment, are different groups disproportionately exposed to environmental hazards? Are schools crumbling in one neighborhood and full of new amenities in another? Laws, incentives, corporate practices—these are all part of systemic structures, and they affect everything from the length of voting lines in different neighborhoods to the likelihood of receiving credit to start a new business or a promotion at work and so much more.

Because system structure is shaped by a system's past (remember path dependence?), equity in the present is also influenced by equity in the past. Whether groups of people have equitable levels of wealth and property in the present is heavily influenced by the system's history. The quality of amenities from one neighborhood to another, the relative wealth of nations, even the burden of toxic pollutants stored in people's bodies (literally called the body burden[1]) are all places where the traces of past inequities manifest today.

Equity, like other system properties, can be maintained or changed by the visions, values, and simple rules of the people within them. As Nathaniel Smith, founder of the Partnership for Southern Equity, often says, the policies in place reflect the values of the people in power.[2]

For example, a recent "resume audit" study on bias involved presenting reviewers with resumes that were identical except for the names at the top and asking them to rate the applicants on various criteria. Those with names that signaled "female" or "non-white" were consistently rated lower despite having similar qualifications.[3] Can you think of a few, perhaps unconscious, simple rules that might explain these results? How might different simple rules produce different results?

Remember also that visions, values, and simple rules derive from worldviews. Let's look again at a few of the assumptions of the Collection of Objects worldview, familiar to us in forms of oppression, including patriarchy, white supremacy, ableism, religious persecution, and more:

- Safety comes from domination.
- The world is organized as a hierarchy with able-bodied, straight, rich, white men at the top, white women below them, people of color still further below, and nonhuman species even further below them.
- Power is the ability to exert force over others.
- Power comes from one's place in the hierarchy.

Since such worldviews shape many systems throughout the world, it is no wonder that inequity exists along so many axes. It's not just built into stocks and flows; it's built into the beliefs and assumptions that influence how those stocks and flows are set up and operate.

Equity is a system behavior like any other. If you want to change CO_2 emissions, or the literacy rate, or the energy efficiency of school buses, you have to change at least one element on our familiar list: structure, vision, values, simple rules, or worldviews. The same is true for equity.

Viewing equity as a system behavior with structural causes will help inform your multisolving. In fact, improving equity is often multisolving in and of itself. For example, inequity leads to health disparities; with reduced inequity, health outcomes improve. But inequity also leads to barriers to education and economic opportunity. With inequity contributing to a set of problems, boosting equity contributes to a whole host of solutions. In her book *The Sum of Us*, Heather McGhee writes about the "solidarity dividend" within which equitable systems promote public health, vibrant infrastructure, economic vitality, and innovation.[4] On a podcast with Ezra Klein, McGhee explained some of the staggering costs of inequity: "You get reports coming out every six months it seems, [the latest one] from Citigroup this summer saying that we've lost nearly $20 trillion in economic output because of the racial-economic divides over the past 20 years. Or the Federal Reserve Bank just this past week—a report saying we lost $2 trillion in economic output in 2019 because of the economic gaps between white men and everybody else. It's no way to run a country. We're leaving some of our best players on the sidelines."[5]

Environmental injustice subjects some communities to hazardous living conditions that might include exposure to toxins and pollutants, lack of access to safe physical activity, extreme heat, flooding, lack of access to affordable, fresh fruits and vegetables, and more. A series of health risks ripples outward from this one form of inequity. The impacts don't stop with health, though. Poor health creates obstacles to education or employment, which can ripple outward to limit economic security. Changing the patterns of policy and disinvestment that contribute to environmental injustice would address these many problems at once. It would be, in other words, multisolving.

Attention to equity will also help your multisolving projects avoid doing unintended harm. Here's an example. Parks and green space are a great multisolving solution. They provide beauty and recreation, help clean the air and cool temperatures, and absorb stormwater to help combat flooding. But they also can contribute to green gentrification.[6] Public funds improve green space with a goal of improving well-being in the neighborhood. But such amenities also tend to make neighborhoods more attractive to new-comers, which can lead to escalating property values and rising taxes and rents, until suddenly the neighborhood's original neighbors can't afford to live there anymore.

A multisolving solution that ends up benefiting the well-off incomers and burdening the neighborhood's original residents isn't really multisolving, is it?

In this situation, a systems view and an equity lens can combine to suggest potential additions to your strategy. Systems thinking tells you to consider the momentum of the system and the feedback loops that could drive such change. When it comes to gentrification there's a reinforcing feedback loop to consider. As housing costs rise in response to the new amenity, legacy res-idents are replaced by newcomers. The newcomers bring more wealth and political connections, bringing even more investment to the neighborhood, and raising housing costs even more.

Prioritizing equity means anticipating and minimizing this reinforcing feedback loop. That can be accomplished in several ways. Legacy residents can be supported to keep their homes with financial and legal assistance, slowing the influx of new homebuyers. Affordable housing projects can be launched in anticipation of the new amenity's construction, adding to the stock of housing in the neighborhood. With more housing available the supply and demand cycle that leads to rapid "overheating" of the market is minimized. Nonprofit structures such as neighborhood land trusts can be established where the price of homes is set not by the market but by the legal structures of the trust. Each of these equity-promoting inter-ventions weakens and slows the reinforcing feedback loop that drives gentrification.

Attention to equity can also help you maximize your multisolving by finding additional benefits. Perhaps you began multisolving to restore a degraded forest. You started off with biodiversity, water quality, and climate goals. Great! But you might, by adding a job training component, be able to address economic inequity in your region. You've identified a way to give even more benefit to your community, and you may have found an avenue to more partners and more sources of support.

Or maybe equity itself is the benefit that first leads you to a multisolving approach. You can harness more allies and resources to go after your goal if you can articulate what else will improve in a more equitable world. More people making a livable wage will turn around and invest in their local communities. More people with paid sick days will be able to stay home when they feel ill, and that will improve public health overall. And so on.

You may be developing new modes of agriculture, new models of land use, or new types of sustainable infrastructure. Each decision you make has the potential to improve equity or to worsen it. If your new transportation program will reduce emissions and provide good jobs, that's great. But what if with a slight adjustment you could also increase the mobility of senior citizens or young mothers? Those additional benefits could increase equity. They would also suggest additional partners and allies.

On the other hand, if you aren't good at seeing intersections with equity, your multisolving project risks leaving potential benefits on the table. Equity won't just happen while you are focused on other multisolving goals. You need to plan for it and design for it.

Often, a focus on equity is also required for multisolving—on any topic— to be effective. The Collection of Objects worldview contributes to many of the problems that we want to address with multisolving. That worldview generates systems that oppress people. Oppression creates sacrifice zones, where the suffering of people and places isn't registered as "mattering" by the system as a whole. This is a weakening of balancing feedback. We saw in chapter 5 that systems can't meet goals for health or stability without vibrant balancing feedback loops.

Many multisolving projects exist to undo this damage, to make the air or water cleaner, to provide access to healthy food or good housing to more people. Therefore, they must grapple with why people's basic needs aren't being met in the first place. Often this grappling will lead to the systemic structures created by these worldviews. You may start with a goal for air or water or health, but your explorations may eventually lead you to understand that the root cause can be found in inequity and the worldview that permits it.

The Collection of Objects worldview can also stand in the way of effective multisolving partnerships. Many of us, despite our goodwill and lofty intentions, still arrive at our multisolving collaborations carrying at least some of the Collection of Objects worldview with us. It would be surprising if we did not, given how prevalent it is in the systems we move through. Yet the habits, values, and simple rules inculcated in us by that worldview are obstacles we must overcome to multisolve well. Successful multisolving requires true partnership, mutual learning, sharing of resources and opportunities.

It requires collaboration across the divisions of our societies. This includes divisions of gender, race, class, and ethnicity. To do a good job multisolving we have to find the Web of Relationships worldview within ourselves and do our work accordingly.

Part of the power of multisolving comes from assembling coalitions with more collective power than any single interest would have on its own. But strong, long-lasting collaborations require more than a common problem. Multisolving partnerships need to be more than transactional, temporary alliances for political expediency. Ideally, they should represent the whole system coming together to care for the whole system.

Solidarity asks something of each multisolving partner. It asks that they care as much about each problem the partnership addresses as they do about what first brought them to the table. For some members of a multisolving partnership, that most important issue may be inequity. It may be the way that burdens are impacting one population disproportionately. It may be the way that benefits are being concentrated in some hands over others.

Not all the members of a multisolving project will always be directly and personally impacted by inequity. But some of them probably will. Multisolving asks that even those who are not directly impacted by inequity make it their problem.

In a multisolving partnership, water advocates may be asked to care about air quality. Jobs experts may be asked to care about transportation infrastructure. And all of them might be called to take a stand for racial, gender, economic, or other forms of justice. Each partner has a role to play in changing the systemic structures that give rise to inequity.

Multisolving projects benefit from the collective intelligence of the participants. No one partner understands the full system, but together their understanding is expansive. Imagine a project to better manage an urban waterway. Some experts understand how water systems work, others understand housing policy. Ecologists know about the potential to improve the biodiversity of the river corridor. Neighborhood residents know what it's like to live in the housing under discussion. They know what it is like to have the stream overflow its banks into their basements.

That's a lot of group knowledge, but without care, inequities that exist between groups in the wider society can limit the potential for collective intelligence. The female ecologist might commonly find herself interrupted by male colleagues. The Black grandmother who has lived beside the river her whole life might have little formal power in city decision-making. Will her voice be taken seriously? Will the group's decisions reflect her lived experience?

Knowledge isn't the only thing that flows between partners in effective multisolving work. Money, resources, and introductions to influential people flow too. The ability to solve multiple problems is tied to these flows. If resources don't flow, or if they flow only in the directions that are typical in the wider, inequitable society, then the potential for multisolving is diminished.

Does the young community leader receive an invitation to pitch a new idea to the project's biggest donor? Is the director of the tiny nonprofit on a shoestring budget given any financial support for her participation in the project? Do the findings of citizen scientists make it into the glossy project newsletter?

Unless multisolving project partners work to change the patterns from the default in the wider society, their efforts can fall short of their potential. Systems don't shift easily or meekly. They push back. They resist change. Multisolving networks need to hold together in the face of this resistance to change. Strong and trusting relationships are part of the glue holding partnerships together in the face of turbulence. And true relationships are rooted in solidarity and care. They emerge between people and organizations committed to one another's well-being and clear about one another's value. Those sorts of relationships are hard to cultivate without challenging systems of domination. For that reason, multisolving efforts will benefit from organizing themselves as webs of equals. Given the power of systems of domination in the wider world, this may require consciously resisting old patterns and shaping roles and responsibilities within the partnership to counterbalance the inequities of the wider society.

Multisolving projects are strongest when they are carried out with a high level of coherence. That includes having a high level of fidelity to the Web of Relationships worldview across all elements of the work.

Imagine your multisolving project is focused on collaboration with nature, an ecosystem restoration protection project that also produces forest-based products. Imagine the project focuses also on economic equity whereby all producers receive fair value for their labor. So far that's coherent and aligned with a Web of Relationships worldview. But imagine the project pays little attention to gender equity. Maybe women are excluded from training opportunities because childcare isn't offered. Or women are discriminated against in accessing credit, but your project doesn't reckon with this obstacle in its design. Not sounding so coherent, is it? The project would have one foot in a Web of Relationships worldview and one in a Collection of Objects worldview.

That type of incoherence has subtle and not so subtle impacts on the effectiveness of projects. It impacts how they feel to participate in. Incoherence doesn't feel good, even when it is unintentional. Incoherence on equity can impact a project's moral standing; maybe the press would rather give front-page coverage to a more equitable project. Maybe, sensing the incoherence, fewer people will want to participate in the project.

On the other hand, a deeply coherent multisolving project with equity at its center can be incredibly powerful. It can see a wider range of possible benefits and avoid a wider range of possible downsides. It can motivate and inspire because it is walking the talk, caring about people *and* land or jobs *and* health.

Equity Practices for Multisolving

If you want to include equity in your multisolving work, what should you do?

Well, to start, you should read analyses of equity that go far beyond what I can cover in one chapter. Read as widely as you can. Read about patriarchy, white supremacy, homophobia, and ableism. By studying different systems of oppression, you more easily see their commonalities. Also read some of the cross-cutting work that looks at the intersections between these systems. Finally, seek out work written by people who are directly disadvantaged by these various systems. Read Black writers on racism, women on feminism, disabled people on ableism. Seeing systemic structures is easiest for the people who experience the harmful impacts of those structures. You can also read about the history of many of these systems to better recognize how they shape current systems.

Beyond reading, how else can you infuse your multisolving experiments with an attention to equity?

Explore your system's current equity picture. Find or collect the data you need to determine the state of equity in the system. What equity gaps are most glaring? Income by gender? Exposure to pollutants or flood risk by race? Energy burden by ethnicity? Access to mobility by disability status?

This information can come in many forms. Often quantitative, statistical data will already be compiled in reports and databases. Qualitative data, such as the lived experience of people subjected to inequities, is also important.

Centering their perspectives can help illuminate the scope and scale of inequity in the system. This quantitative and qualitative information will help you in at least three ways.

It will enable you to understand the most pressing dimensions of inequity to attend to in your project planning and implementation and what partners and perspectives will be needed for an effective project. Accurate information will also show you which dimensions of inequity your project might be at risk of inadvertently replicating. Finally, establishing data on inequity at the start will give you a baseline from which to notice changes. If your work is successful, you'll be able to point to contributions your work may have made to improving equity, and you may discover ways to be even more effective in the future.

Develop your own project equity metrics. You've worked to understand the dimensions of equity that are most important or prominent in the systems you are trying to influence. Now, you and your partners can think about how those inequities might reveal themselves in the structure of your project.

Say you've identified inequities that mean women from low-wealth communities are particularly disadvantaged. That's a dimension of equity you might want to measure in your project and reflect on your internal design. What might you measure? Perhaps you'd want to know what percentage of your multisolving collaborative is represented by women from low-wealth groups. If the percentage is smaller than it is in the community, you might want to take corrective action.

You might also want to track "airtime." Are the conditions in your meeting such that those women speak and give as many presentations as people from other backgrounds? What about financial resources? Maybe your collaboration disburses small grants to participants in the collaborative. How does the share of grants to this group compare to other groups? What about attribution and recognition? When opportunities for external exposure arise, are their names put forth? Merely tracking equity metrics doesn't change systemic structures within your project, but it will alert you if you are mirroring inequities in the wider society, which gives you a chance to explore and address potential causes.

You don't have to stop at measuring the equity of your system (in this case, your multisolving project). You can steer the system toward greater equity. Remember: structure creates behavior. Equity is an emergent property of a system. A system's equity depends on its elements and how they are connected. That systemic structure depends on the visions, values, and simple rules of the people who influence the system. And those visions, values, and simple rules are shaped by the worldview that underlies them. All of this means that, both within your project and in the wider world, you can change the structures that create inequity.

You may need to change the levels of stocks or the rates of flows. You may need to create new feedback loops or interrupt existing ones. You may need to strengthen some visions and values and turn away from others. Luckily, there are many ways to do this.

Reduce obstacles to participation. Structures in inequitable systems put more obstacles in the way of some groups of people than others. If you don't look for these obstacles and find ways around them, those obstacles will weaken your multisolving efforts. If you hold your meetings during working hours, you might attract professionals but make it hard for volunteers who have other "day jobs." If you don't offer childcare, young parents might not be able to participate. If your meeting is far from transit stations, only people who own and drive cars can easily participate. To break down silos in the world you will also need to break down barriers to participation in your efforts.

Develop intentional, shared vision, values, and simple rules. Both you and your partners will likely bring unconscious biases into your joint work. How could you not, coming from inequitable societies? Unconscious bias is, in essence, an unexamined simple rule. *Pay attention when an older white man talks. Discount presentations that don't look "professional."* And so on. As simple rules, biases influence how groups of people come together and what they can accomplish.

Vision, too—or rather, lack of vision—can pose limitations. Some of your partners may have been taught to accept that the status quo is "just the way it is." This resignation risks limiting your vision of the full possibilities for your work that would benefit more from an expansive vision of an equitable society.

As for values, who and what matters? What do we owe each other? What do we owe nature? What are our obligations to future generations or the most marginalized? What does leadership look like when it comes from those who have historically been excluded?

Taking the time to discuss visions, values, and simple rules will help you identify the unconscious ones that might not support the spirit of the work you hope to do. Naming the values to which you do subscribe is important too. In moments of confusion, sudden change, rising opportunity, or new threats, your shared values—especially if they have been articulated in advance—can point the way.

Experiment and learn. If you are trying to create the conditions for equity but you and your partners are used to systems that aren't equitable, you'll need to experiment with new ways of being and working. In this regard many of the skills of operating in complex systems will put you in good stead. Return to the section on cultivating a systems stance in chapter 7. How might those practices help disrupt habits and systems structures that no longer serve?

Try shifting roles. Have the people with the least formal power helm decision-making. Have the person who leads the largest organization take the notes or fetch the coffee. Those who are usually quick to talk can step back and listen. Perennial teachers can recast themselves as learners.

Infuse everything with learning opportunities. Make the giving and receiving of feedback (even uncomfortable feedback) normal. What's working? Why did that feel so bad, and what can we do better next time? It's going to take learning, discomfort, and vulnerability to shift systems into more equitable forms. There's no handbook and no predetermined path. Stay open, commit to learning, and you may be surprised by what you accomplish and where the journey takes you!

Creating conditions that support equity within a world that does the opposite isn't easy or comfortable. For people and organizations that are advantaged in status quo systems, this work comes with stepping back. You might be asked to listen when you are accustomed to being listened to. You might need to release resources, money, power, or relationships that you are used to thinking of as "yours."

As a middle-class white person with a lot of formal education and an organizational leadership position, I try to push myself to spend more time in a "stepping back and listening" role and continue to ask, "What resources can I share or release?" And I have more to learn about all of this; it will be a lifetime's work.

For people from marginalized groups, stepping forward may be what helps shake up inequitable systems. As a woman working in fields that are often dominated by men, I have a small taste of this role too. I've had to work hard to value my own voice and perspective and to offer it fully.

As an antidote to the frustrations and challenges on both sides, I suggest trying to keep the big picture goals in mind. We can't multisolve well from within the roles the Collection of Objects worldview assigns us, and in our time of converging crises I don't see other alternatives to multisolving, painful and slow though it may sometimes be. If we need each other to make it safely through these times, then we need to keep seeking understanding, root cause solutions, and deep structural changes. There's no simple substitute, at least none that I can see.

Remember our experiences of inequitable systems are not the same. Dr. Camara Jones is a past president of the American Public Health Association and a masterful user of parables to elucidate points about racial equity from her perspective as a Black woman, a physician, and a public health expert. Here's a description of one of her parables, based on a keynote she gave in Chicago in 2018 that captures an important characteristic of inequitable systems.

> As a parable of how the reality occupied by people of color is typically invisible through the lens of white privilege, Jones offered an analogy from her days as a medical student. She and her friends had been up late studying when they decided to go to a nearby diner for a late-night meal. They were seated at a table and served, but later she happened to notice the sign on the door. From where she sat, she could see a sign that read "Open," yet passersby outside would stop at the door as if they wanted to enter but then would move on.
>
> Clearly, the diner had closed to more customers since she and her companions had been seated, and what looked to her like a sign saying "Open" was a sign saying "Closed" to those outside. "Racism structures a dual reality," she remarked. "Those inside the restaurant sitting at the table of opportunity don't even know that there is a two-sided sign.[7]

It will serve you to keep in mind the "dual reality" Jones describes while multisolving. Which side of the sign are you used to seeing? And what do your colleagues and partners in the work experience as they move through the world?

No decisions about us without us. That's a slogan you see on posters at grassroots protests about environmental justice. I learned recently South African disability activists made the phrase well known in the 1990s.[8] James Charlton, author of a book with that title, reports that the South Africans in turn adopted the phrase from Eastern Europeans they met at an international disability conference.[9] Sounds like the sort of "sticky" concept we discussed in chapter 4, doesn't it, spreading easily by word of mouth, because it captures something important? It's also a good reminder about your multisolving project composition. If you ever reach the point where you realize you're talking about policies or investments that will impact groups of people who aren't involved in your project, that's the sign that you need to expand the bounds of your project. You need to add to the collective intelligence.

Those are some inward-facing aspects of bringing equity into your multisolving projects and partnerships. Applying an equity lens to the larger systems you are working to steer is equally important.

Ask, "Where do we have the power to change how resources flow?" Think back to our discussion of stocks in chapter 2. Do you recall the image of the observer, high at the top of a watershed, looking out across a landscape of ponds and streams? You and your partners can seek out this kind of perspective from time to time. Where are resources accumulated? Where are they needed? Are those the same? If not, what flows would bring the system to a healthier balance? Who controls those flows? Who could shift them? Which stocks are always filling, driven by the way those with power have structured the rules by which the stocks flow and interconnect? Which are draining, their vital resources flowing off elsewhere in the system? Ask where you have the power to change how resources flow.

You can explore the system's history and path dependence, a history that some in your multisolving network may know all too well, while others never learned much about it. Building a shared map of the structures—both as they are and as they could be—can improve your conversations and your strategy.

Ask, "Who benefits and who's burdened?" For whatever intervention you are planning you can ask how benefits and burdens will be shared. Will renters be able to access the future cost savings of rooftop solar panels, or only homeowners? Will the boom in battery storage for the clean energy economy create toxic waste for some other community? If potential exists for burdens to fall on already marginalized groups or benefits to accumulate for the already privileged, that's a sign your effort needs to give more thought to project design.

Remember the principle "no decisions about us without us?" That will make it easier to do a thorough job answering the "who benefits and who is burdened" question.

Disrupt "success to the successful" reinforcing feedback loops. "Success to the successful" is the name system theorists give to a particular reinforcing feedback loop that sustains inequity.[10] Like its name suggests, this loop creates a behavior where the more of something—such as wealth or power—one part of a system has, the more it's likely to get. For instance, an already powerful political party might enact laws explicitly to give itself more power and thus even more influence on rulemaking. Or a wealthy institution might use that wealth in new ventures that allow it to acquire even more wealth.

Success to the successful loops, left unchecked, generate inequity. Policies that promote sharing and redistribution are the way to disrupt "success to the successful." That might look like taxing the wealth of the very wealthy in order to fund projects that benefit the common good. It might look like opening up opportunities in job training, education to people from marginalized groups. It might look like reparations for past harms that also concentrated wealth in some hands, or it might look like the redistribution of access to land. It might look like ensuring voting rights and getting money out of politics so that those with wealth and power don't have more influence on governance than anyone else.

It's also worth looking carefully at benefits that your multisolving project might create. Look for any inadvertent potential for success to the successful. How can early beneficiaries of your efforts give back so that others coming along behind them benefit too?

———⋘———

As the word *multisolving* has moved into more and more common usage, I have noticed that not everyone emphasizes that multisolving is an approach rooted in equity or insists on the relationship-based, nontransactional elements of multisolving. So, here I'd like to clarify a few things that are *not* multisolving as I see it.

It's not multisolving to talk about certain benefits of a solution in order to avoid discussion about controversies or burdens. That's a political calculation, not multisolving. For instance, it's common in some circles to avoid naming the climate emergency. Instead, some political leaders tend to speak only to the jobs that could be created by building clean energy infrastructure.

That may or not be an efficient messaging strategy. What it's not is multisolving. Multisolving acknowledges every piece of the system and develops strategy with them all in mind.

It's also not multisolving when additional issues are tacked on in a temporary way for political expediency. Multisolvers don't add a constituency to their coalition to win an election only to walk away from them in the aftermath. Because multisolving doesn't elevate some goals over others, true multisolving efforts won't "sell out" those who care about one problem to make progress on a different one.

? Questions for Reflection

- What are the biggest equity gaps in the systems that you live and work within?
- Do you think multisolving is possible without a commitment to equity?
- Have you ever experienced "success to the successful" either as the beneficiary or as someone who was disadvantaged by it? Describe the situation. How did it feel?
- In the systems you live and work with, what problems ripple outward from inequity? What benefits would be produced by increasing equity?

The Lights of Shore

The wind whips
from the north, then the west,
from every direction.
The little boat spins.
Waves crest and dip.
We are spinning, swirling,
wind tossed.
Tiny,
a toy boat on a planetary sea.
All our treasures are stowed here
on this creaking boat of
     battered wood.

We are spinning again in
     the lashing rain.
You stand so brave in the prow,
eyes on distant lights.
The wind dies down.
Everything quiets.
For how long?
I pull on the oars,
You point the way.

CHAPTER ELEVEN

Multisolving in Tumultuous Times

We want our multisolving efforts to last and thrive, even if conditions change. Having put in all the work of identifying changes that improve multiple problems at the same time—and building the needed partnerships to make and sustain those changes—it would be tragic if some other crisis just beyond the boundaries of the project's focus were to destabilize the effort and erode the gains. That's why the best multisolving efforts design for potential destabilization wherever they can.

There are a lot of factors that could make the coming times tumultuous. There are possible climate change impacts. There are the social, economic, and technological changes that will be needed to prevent the worst impacts of climate change. While welcome and needed, some of these changes can still bring stress and strain, and some of them could impact the most well-thought-out and innovative multisolving projects. Then there are ecological crises and economic shocks to come.

The good news is that a systems perspective can help you think about designing your multisolving projects so that they can thrive even in tumultuous times.

This chapter builds on everything we've covered thus far—stocks, flows, feedback loops, equity, vision, values, and more—and provides considerations to help bolster the flexibility and resilience of your multisolving efforts so that they can best meet unpredictable or even destabilizing conditions.

A necessary caveat, and one that can't be repeated enough: there are limits to the magnitude of shocks we can prepare for or withstand. We can make cities cooler, but if temperatures and humidity rise above certain levels, it becomes dangerous for people to be outdoors for much time at all. We can plan for stronger rainstorms, but if a month's worth of rain falls in a few hours, even really good planning will be overwhelmed.

Your multisolving may need to happen in the midst of crisis. It may need to adapt to changing conditions. But it cannot *only* be about adapting. It must *also* contribute to getting to the roots of crises. That's the only way to make sure that symptoms remain manageable.

Prepare for Shocks

In systems parlance, a shock is a sudden change in an element of the system. A sudden change in a rate is shock, like when a drought strikes and the average rainfall per month plummets. A supply chain disruption is a shock that happens when some flow along the chain falls or stops entirely. Supply chain shocks rippled through the global economy when COVID-19 outbreaks struck factories or ports. These disruptions offer an example of another feature of shocks—they ripple. Production of microchips needed to assemble new automobiles dropped, which slowed production of new cars. That shock rippled into the used car market, where prices suddenly rose.[1] It doesn't take much imagination to see those ripples reaching into your multisolving project: your project manager needs a new car, is forced to spend more than she budgeted, is stressed about the payments, or takes a second job to make ends meet. Or she doesn't buy the car and is late for work because the bus is unreliable. Or . . .

One of the most obvious sources of shocks to come are the climate impacts that we can't prevent because of warming that has already happened or warming that will come from fossil fuel infrastructure that has already been built. Some climate shocks will be felt very locally—specific floods, fires, heat waves, and droughts. With better and better climate change modeling, you can have an informed understanding about which types of impacts are most likely for where you live.

Other impacts of climate change will be of the "ripple" variety. A drought in a place where a staple crop is grown could raise food prices halfway around the world. People relocating from coastal areas in response to sea level rise could increase the cost of rent inland. And so on.

Yet other "shocks" could come from the measures needed to limit climate change in the first place. Limiting new fossil fuel extraction is a needed policy to protect climate and health. But in communities where that extraction is a source of livelihoods, the change could be experienced as a shock. A new protected bike lane in a city could be a win for climate and health, but it might disrupt local businesses along the route during the period of construction. In the decades to come, some of the disruption we face will be badly needed, but it will still be disruption!

When preparing for shocks, we need to think not only about direct impacts and sudden shocks and the ripples they create. Even if a part of a system is changing in a steady, predictable way, the existence of thresholds within systems can lead to a steady change being experienced as a sudden shock.

A stock may fall and fall and fall (or rise and rise) without destabilizing the rest of the system, but once it falls below a certain threshold, the system could suddenly shift into a new behavior mode.

Shocks impact multisolving projects in all sorts of ways. What if your program to weatherize hundreds of homes can't obtain needed building materials because of supply chain disruptions (or skyrocketing material costs)? What if the power is out for weeks after a storm and you can't communicate with partners? What if a new ruling on clean energy bumps up the price of electricity for your senior center? The shocks to which any particular multisolving project are most vulnerable will be different, but the potential for disruption looms for all of them.

Knowing that shocks from climate change, biodiversity loss, rapid decarbonization, and economic destabilization are all potential parts of the landscape within which we will be multisolving, let's look at some ways of bolstering your work against shocks.

Maintain key stocks at high enough levels. As discussed in chapter 2, stocks help buffer against shocks. You can take advantage of this property in your multisolving projects. In the systems where you are working to support multiple goals you will notice points of accumulation that help the system stay steady in the face of fluctuation. Making sure these stocks are sufficiently high is like an insurance policy for your multisolving. It will help make sure all your good work can't be wiped out by a disturbance to the system. This is most obvious for stocks of essentials like food and water or spare parts for critical systems, like water supplies or power plants, but it applies also to stocks of knowledge, coping capacity, personal energy, and collective goodwill.

Think through the potential disruptions or shocks that seem most likely in the context of the system within which you are multisolving. Consider what the system would need to last through that disruption in health and safety. Compare that to your actual stocks. Do local shops have enough food for a week? Does the new youth center have enough cash reserves for a year?

Remember that not all stocks that matter are physical ones. How is your town's stock of people who know how to fix the water supply? How is the community center's stock of people who know how to generate paychecks when the computer is offline?

Keep some stocks low. Other stocks provide buffering and resistance to shocks when they are empty, not full! I worked on a project for a few years in the Milwaukee area where the sewerage district maintains what they call the Deep Tunnel, a hollowed-out rock formation with the capacity to hold millions of gallons of sewage and stormwater before it reaches wastewater treatment facilities.[2] If the tunnel were overfull, the treatment plant could not keep up with the combined flow of storm- and rainwater during intense rains. Untreated wastewater would spill into Lake Michigan, creating an environmental and health hazard. In the case of this tunnel, keeping the stock empty creates resilience.

Think about the systems you are engaged in. What stocks must be kept low to buffer against potential shocks? Do you have a way to compost food scraps if the garbage truck drivers go on strike, or will smelly food be accumulating in your overflowing trash can? Does your worker co-op have enough space to store product if the trucking company that carries away product for distribution is disrupted?

Invest in equity, so that everyone can be resilient. It's easy to suggest keeping key stocks at the right level to be ready for shocks. But of course, inequity, including economic inequity, can get in the way.

Workers who live paycheck to paycheck may not be able to afford weeks' worth of supplies of food in the face of hazards like hurricanes or wildfires. Their homes may not be large enough to store lots of extra food. They may lack a home altogether. A hospital system in a marginalized community may not have the cash flow to stockpile medicine and supplies. The whole system could be more vulnerable if a supply chain disruption, storm, or power outage made resupplying difficult. Restructuring systems to allow for higher levels of equity can be a prerequisite to helping them prepare for shocks.

Pay attention to the capacity to refill (or drain) critical stocks. For a short-term shock it may be enough to set aside critical supplies or to make sure that your capacity to hold outflows like products or wastes is sufficiently large. But for shocks that might extend longer than your supplies (or holding capacity) might last, you'll need to focus not just on the stock itself but on the capacity to replenish (or drain) it. Ask not only how many dozen eggs you should have but also how many chickens; not only how many gallons of clean water but also how many pumps and purifiers to provide new supplies.

Finally, don't forget stocks of knowledge and skill, which are necessary to manage other stocks. Having enough gardening tools and seeds in advance of a food shortage but not having experimented with gardening would not be ideal. Your ability to replenish the stock of your food supply depends not only on the stock of equipment but also that of knowledge.

Prepare yourself for flows that might fluctuate unexpectedly. Wherever you can, try to anticipate some of the places where fluctuations in flows could undermine your progress.

In the fall of 2020, Senegal received as much rain in a single day as it typically does during the entire rainy season, resulting in the city of Dakar flooding and thousands of people losing their homes.[3] That's an extreme example of tumultuous times. Around the world the frequency of these sorts of extreme events has been rising and is predicted to rise even more with climate change. It's also an example of a flow suddenly jumping to unprecedented magnitude; few systems, including drainage systems, buildings, and roads, can cope with such large and sudden changes. A multisolving project to support small entrepreneurs in Dakar could have been devastated by these events. It's very challenging for individual projects to design around such extreme events, but they can pressure local and national governments to upgrade infrastructure better suited for changing conditions and create social safety nets to help people and organizations better ride out and recover from shocks. Individual projects may not be able to do all of that on their own, but that doesn't mean there's nothing worth trying. To continue the imaginary example in Dakar, those worker-owned co-ops could try to site themselves above flood level, they could store inventory on upper floors of buildings, and they could plan to spread the project across multiple locations so that not all resources are in a single, potentially vulnerable spot.

Under the influence of converging crises, flows can drop precipitously as well. In 2018 the southwestern United States experienced a major drought as well as an intense fire season.[4] Cattle producers suffered, and in New Mexico surface water supplies dwindled. Flows that are too small can cause as much trouble as those that are too large.

Of course, waterways aren't the only flows that fluctuate in tumultuous times. As businesses falter in economic downturns, sales fall, property values fall, and workers face unemployment. Tax revenue to governments falls as a result. Governments are less able to offer services and supports, even though their constituents need help more than ever. If your multisolving idea depends on this sort of support, then you've identified a potential vulnerability of your project.

Other vulnerabilities may include flows of raw materials or important repair parts. If a flow is critical for you, your community, or your business, consider planning for what you'd do if it suddenly stopped. Conduct an audit of the flows that shape the arena in which you are multisolving. Which ones are most likely to swing outside normal ranges due to climate change, biodiversity loss, economic shocks, or supply chain disruptions?

Anticipating the floods and droughts (both literal and metaphorical) that could be heading your way will increase the odds of your project safely navigating tumultuous times.

Develop strategies to protect vulnerable flows in your multisolving projects. Many of the economic flows that characterize our world today are long and complex.[5] As you build multisolving solutions it's worth minimizing your dependence on critical flows that could be disrupted by events that are distant or out of your control.

Is your community garden project dependent on flows of seeds from distant suppliers? If so, can you begin to include seed saving and sharing in your program design? Does your program to replace gas boilers with heat pumps rely on technologies that are produced in another country? Are there any alternatives? Could local manufacture become a long-term goal of your project?

Look to a Future That's Different from the Past

We know that systems have momentum. The systems we live in today are impacted by the momentum of climate change that we can't prevent and biodiversity loss that is already under way. Global processes like the Paris Climate Agreement as well as national climate targets and policies are trying to build momentum for decarbonization. That momentum is already transforming industry, transportation, cities, and agriculture. In other words, the world your multisolving project will mature into could be very different from the world today. For your project to continue to thrive and contribute you need to design it for the world that is coming, as best you can see it. Here are few strategies to consider.

Build long-lived stocks for the world that is coming, not the one that is here now. To the extent our budget allowed when we bought our land twenty years ago, my eco-village neighbors and I built our homes with green technology. We added efficient windows, thick insulation, composting toilets, and solar water heaters. You could say we were doing our best to multisolve for health, affordability, and low environmental impact.

Now, with the benefit of hindsight, I can see how those original decisions constrained possibilities. Had we invested in slightly thicker insulation for walls and roofs we'd be closer to today's net-zero standards. We might have anticipated the warmer, more humid nights that are now causing mold and mildew problems we didn't experience in those early years. We might have been able to do more to build for the future that was coming rather than the present we were living in.

Cars, power plants, office buildings, trains, and highways are other examples of stocks whose long lifetimes mean they may persist into a future that's very different from the present. News is already being reported about heat waves disrupting air travel. On extremely hot days heavy planes need longer runways, and safety calculations may need to be recalibrated for unusually high temperatures.[6] Airports and planes were designed with the expectation that future conditions would continue to be much like the past.

How long will a given stock be around? What do we know about the future conditions that it will exist within? These are good questions for those involved in multisolving efforts to keep in mind.

Spread and slow spiking flows. For flows that might grow suddenly larger, see if you can anticipate that possibility in your plans and designs. For instance, many towns in my region are systematically replacing drainage culverts with larger ones, anticipating climate change's more severe precipitation in the northeastern United States.

In other parts of the country, communities are also investing in rain gardens, green roofs, and wetland restoration. Each of these interventions captures the extreme flows. The gardens and wetlands hold the water and slow it down, rendering it much less destructive.

Finding ways to slow and spread suddenly large flows applies to more than water. What if a country is expecting a sudden influx of new citizens, say climate refugees? How could a nation respond to a concentrated flow at ports of entry and distribute that flow so it is less of a shock? Can you imagine systems designed to help new arrivals settle across the country? Can you picture ways to distribute the effort across networks of supporters so that no one agency or community is overwhelmed?

Where can you make plans to handle bigger flows than you currently experience? Wherever you can, devise ways to spread out and disperse intense, heavy flows.

Cultivate redundancy, diversity, and simplicity. Many mainstream systems meet a narrow definition of efficiency. They are optimized to produce the most product for the least cost. That might be a winning proposition during a period of stability. But what happens during a period of rapid change?

It might be efficient to maintain a small inventory of spare parts. After all, that inventory represents money that could be invested in some other aspect of productivity. But what happens with a disruption to the supply chain for that part? Building in a larger supply of surplus parts could be protective.

A specialized manufacturing facility might be efficient and profitable during stable times. But what happens in a crisis? At the beginning of the COVID-19 lockdown there were shortages of toilet paper. The factories that made toilet paper for office buildings struggled to retool to make toilet paper sized for the rolls used in home bathrooms.[7] Society wasn't suddenly using more toilet paper. Demand had simply shifted to a different shape. But since it was highly specialized, the system was unable to quickly adapt to the new need.

We see something similar with monocultures in industrial agriculture. Planting thousands of acres in the same crop can be very efficient. Growers can specialize and don't need to maintain a variety of equipment. But the practice can make the crops vulnerable to pests. A more diverse farm growing a variety of crops might be less efficient but would be more resistant to the disruption of a pest outbreak. Even if some crops were impacted, others might still thrive. Diversity, while not always efficient in a narrow sense, can be protective in times of disruption or rapid change.

These examples all highlight a trade-off. Efficiency can help systems thrive during periods of stability. But it can make them vulnerable during periods of disruption, hindering their ability to shift in response to new opportunities. You are already making progress in this shift by multisolving. By designing for multiple goals, you are creating a system that is less highly optimized for singular goals and therefore likely to be less brittle.

Multisolving often happens in partnerships and collaboratives and thus can help build redundancy and flexibility. You can have more than one organization maintaining the key records or holding knowledge on key processes. Since you are already experimental and iterative in your design, you can try several ways to meet the same goals. Not only will you learn, you'll also have backup plans and approaches should you need them.

Retreat. In the face of destabilization that can't be coped with, there will be times that the best course of action will be retreat. Not all the places currently occupied by humans will be places that we can thrive or even survive. Even with building up protective buffers, planning for new conditions, and designing for redundancy and diversity, there will be communities inundated by the sea, in constant danger from wildfires, or without access to adequate water. Political or economic instability might put your multisolving projects at risk. Whether literally moving to higher ground or shifting priorities or sectors, there may be times when the wisest course of action is to leave. Much has been written on this topic, which is often called "managed retreat."[8] It's a tool that will be a part of multisolving during tumultuous times. Managed retreat, where it is called for, won't be easy, but it will work better with an awareness of systems, complexity, equity, and multisolving.

Expect Opportunities

Not all shocks are dangerous, and even dangerous or harmful shocks can set off ripples that are positive or that present opportunities. For example, a climate disaster, even while devastating, could contribute to a spike in new climate activists, which could vitalize a movement and help lead to a system less prone to future climate shocks. Obviously, there's nothing to celebrate in disaster. But being prepared for coming shocks and destabilization also means being prepared for increased urgency, openness to new ideas, and rising readiness to act. Your multisolving projects can anticipate such moments even if you can't predict precisely when they will happen.

Remember also that thresholds and tipping points can produce motive force in the direction of changes you are working toward. Movements that were once on the margins can cross thresholds of relevance where they wield influence at centers of power. Technologies that were once too expensive to scale up can suddenly displace the reigning dominant technologies and change the landscape of a whole industry.

If you want your multisolving to be prepared for opportunities that could arise with turbulent times and discontinuous change, here are a few strategies to consider.

Expect surprises, including pleasant ones. You've planned well. You've attended to every link of a reinforcing feedback loop. You've grown the power to change systems or spread a new idea. That's great, but it doesn't mean you won't be surprised.

When change feeds on itself, small variations in initial conditions can create large differences in the final results. You can build recognition of these sources of unpredictability into your strategies. Plan, if you can, for the upside. Imagine your feedback loop takes off beyond your wildest dreams. What will you do to handle the new cash, the new supporters, or the new opportunities?

Also, design your strategies with multiple seeds and multiple reinforcing feedback loops. In a world where not all seeds grow to fruition, planting a lot of them is sometimes the best strategy. Along with planting a lot of seeds, cultivate a learning approach. Say you are promoting a new idea via reinforcing feedback. Don't just design one campaign. Design six or eight. Test different messages, with genuine curiosity about which ones are stickiest. Create a strategy to collect data so that you can compare across strategies and learn for future campaigns.

Emergence brings its own flavor of unpredictability too. You've nurtured relationships and created connections. Silos have been bridged and webs of knowledge and information flows created. That is great as well, but that doesn't mean you won't be surprised! When change emerges from interconnections, the result can be something you wouldn't have expected from the behavior of the parts. What does expecting surprises look like in this regard? Cultivating a keen sense of observation helps: What is changing in the system? What new behavior modes are arising? What is possible that wasn't before? What is enabling the changes, and can you support those enabling factors even more?

Recognize that reinforcing feedback loops aren't always controllable. It won't always happen, but sometimes your strategies based on reinforcing feedback will grow beyond your capacity to control them. The creator of a new idea may be able to see only a part of what it means or could enable. The creativity of others might alter the idea, translate it into new contexts, or apply it to new problems. The idea may even change to mean different things than what the original creator had in mind.

In a complex world we can only see the future in shadows, in potentialities and scenarios. If we are tapping reinforcing feedback, this is even more true. Holding on too tightly to control can limit the full possibilities of the seeds that you plant.

Prioritize Care

You are a system moving through tumultuous times. Your organization is a system challenged by disruptions (and maybe also by big opportunities). Your neighborhood (or nation) is trying to navigate as a system in a changed world. All of these will benefit from bolstering support, compassion, and care. Here are a few ways to do that, and you can likely think of more yourself.

Remember your "flow nature." In tumultuous times the unexpected keeps happening. Shocks, setbacks, and leaps forward start to feel more common than slow and steady progress in a particular direction. In such times it's good to remember that you yourself and everyone you love are not static.

You are made of moving, flowing parts, which themselves may also be made of moving, flowing parts. You grow and shrink, steady and bend, shift and migrate, learn and experience. You change all the time. You are change. Who better to face a time of rapid change than someone who has been changing their whole life?

Remember also that the networks you are cultivating and participating in have flow natures. New members join and sometimes old members move on. With changes come new experiences, personalities, and webs of relationships to draw on.

Carry with you the image of yourself and your networks not as billiard balls bouncing across a rough and unforgiving and increasingly bumpy landscape but more like water. You pool to rest in a quiet spot. Then you rush energetically through rapids that open to a broad and shallow river until the next stretch of rapids appears.

All of those systems that seem so threatening right now? The gigantic military-industrial complex? The fossil fuel industry? The structural racism that permeates some systems? Each is, just like you, nothing more than stocks that are constantly flowing. Starve their inflows and they will wither away. Boost their outflows and they will fade from significance.

Since you can't keep a world of flows from changing, even if you wanted to, embrace the flow nature of life instead! Build a deep and intimate knowledge of flows and base your work and your strategies on it.

Attend to the balancing feedback loops within yourself. In times of crisis and rapid change, some of the most important balancing feedback loops to nurture are the ones within ourselves. When we get tired, we need to rest. When we are afraid, we need support to feel safe. When we are confused, we need ways to make sense of new information. When we feel small and alone, we need community and support. After all, being healthy, steady, balanced, and regulated is a prerequisite for doing anything else within systems.

Setting your own goals for the state of the system called "you" is part of the work of being effective in complex systems. What does this look like for you? Being well-rested? Nourished? Calm? Reflective? What do you know you need to be at your best?

Learn to recognize the "gaps" of these balancing feedback loops. These are the signs that the system that is you is out of balance. What does it feel like when you aren't well rested? When you're anxious or overwhelmed? Once you are clear on what it feels like to not be at your optimal state of well-being, of course you'll need mechanisms to re-find balance. That is easier said than done sometimes. It can be the work of a lifetime to figure out the care and feeding of your own system. As with most things in complex systems, be experimental and playful. Pay close attention to the results of your experiments.

When you are emotionally drained, what works better: sleep, art, a funny movie, a meal with friends, or a walk in the woods? Try out different balancing loops and see which ones seem to steer you back into the state you will come to recognize as health, balance, and resilience.

Support the balancing feedback loops of living systems and communities. All living systems have balancing feedback loops that are healing and regenerative. That includes individual humans, families, organizations, and communities. It also includes ecosystems and ecological communities. Supporting the self-healing of living systems is a critical skill for multisolving in tumultuous times.

When Tropical Storm Irene devastated my region in Vermont, citizens and organizations of all sorts leaped in to help recover and rebuild. A local brewing company opened its doors for people to pick up water and ice. Churches cooked and delivered meals. A home repair nonprofit tapped its mailing list to become a hub that matched volunteers eager to help with families needing help to clear mud and debris out of flooded buildings. The fabric of the community contained a natural recovery-promoting balancing loop. That loop became active long before federal disaster officials appeared. Those churches and organizations and small businesses and the links between them are all part of my community's own self-healing capacity. Investing in that sort of capacity can be one of the best ways to prepare for tumultuous times.

Ecosystems, soils, and watersheds all have within them health-restoring balancing feedback loops too. Sometimes these are active. Sometimes they are latent and need to be activated or encouraged. Sometimes they need pressures (like overharvesting or development) reduced. Whether it is restoring degraded lands, creating ocean protected zones, or using agriculture practices that promote biodiversity, we can care for natural systems in ways that support their own self-regulating and healing capacities.

The only thing more difficult than having a systems awareness as you move through destabilization and shocks is moving through destabilization and shocks without systems awareness.

Much of current and coming destabilization is not within our control as individuals or even as leaders of substantial projects. We don't get to choose whether or not we have to cope with destabilization, but we do have some control about *how* we cope. We can choose a systems stance of humility and continuous learning. We can choose solidarity and connection rather than isolation and silos. And we can try to anticipate shocks and create the buffers and networks of connection that can help us withstand them.

Most of all, we can always be doing more than one thing at a time. We can navigate tumultuous times while addressing the root cause of that tumult. In fact, that's our only choice. Focus solely on root causes and we risk our efforts being wiped out by a storm or an economic crisis. Focus only on the turbulence and we'll lose the chance to steer systems into more just, equitable, and sustainable behaviors. The narrow zone to aim for is in the middle: focus on addressing root causes of the tumult while surviving (or even thriving) in its midst. In that spirit, let's turn to the final chapter of the book. Let's turn toward going forth.

? Questions for Reflection

- For your own life, community, or organization, what are some stocks that might be most impacted by the kinds of disruptions or shocks that your region is expected to experience in the coming years? How are you (or could you be) prepared for potential shocks?
- What are some flows through your business, your community, or a project you care about that might be subject to sudden and unexpected changes (either increases or decreases)? What could be done so that those sudden changes are harmless (or at least less harmful)?
- How do you cultivate balance in yourself? In the teams you lead or participate in?
- What would it look like to prioritize care? How would you be different? How would the systems around you be different?

Which Way

Will you be the one who turns a
huge boat on rough seas?
Or is it you who gathers up and
comforts the lost children?
Will you find kindness
when fear rises?
Will you know when it is time
to walk away with only
what you can carry?
Is it you who has the courage
to quietly wait?
It's all happening at once now,
the fires of destruction and
the fires of invention.
Inside the chrysalis, change unfolds.
Living rafts, strewn with
flowers, launch.
You are choosing, again and again,
which way to go.
Nothing changes.
Everything changes.
And you are the still point
in the swirl.

You bind together an improbable
convergence of good things.
You undo.
You redo.
Listening, you move.
In your wake everything shifts.
You are never alone.
Wild bees, birch trees, glacier water,
old teachers, new life,
we are all with you, conspiring.
Pass seeds hand to hand.
Make compost.
Braid vines together.
Channel rain.
Speak truth.
Look inside.
Renew.
Grow a taproot.
Know where you stand.
Bend with breeze.
Hang on.
Go forth.
Don't stop.

CHAPTER TWELVE

Going Forth

Here we are in the final chapter of the book. By now you have probably seen enough of how I think about systems to not be expecting the unveiling of a big secret.

I have no list for you of five simple steps to multisolve for the particular cluster of problems you face, nor a well-laid-out roadmap for your terrain and no recipes for exactly what you need to thrive in tumultuous times. I don't plan to offer you a remedy for the feelings of vulnerability or uncertainty of being alive in these times. There's nothing I can give you that will protect you or the people and places you love from harm, or even prevent you from doing unintentional harm.

Already today you can find plenty of books that will give you a formula for these times with a well-polished sense of certainty. They will tell you who to blame or what state or island or mountaintop is the best refuge from coming chaos. You can buy products to keep you safe, bunkers to stock with canned goods, walls, gates, and weapons to protect yourself.

I expect this is the beginning of a trend that will only escalate as the instability and fragility of systems becomes more and more apparent, and that there will be people with all sorts of trademarked products and technologies and methodologies. You'll find people who will tell you all that's needed is to live closer to the land. Others will say safety lies in the direction of high-tech cities. Buzzwords will come and go. There will be a temptation to armor oneself and everything one holds dear, and many purveyors of various sorts of armor. I expect there will be gurus, strongmen, and conmen, and also good faith leaders with good ideas. There will be plenty of purveyors of certainty. Yet everything I know about systems tells me that anyone who seems certain about how this all will end up doesn't know what they're talking about.

As far as I can tell, the path to a future where all life can thrive will be confusing and dangerous. Sometimes it will be terrifying, sometimes elating. It will be both beautiful and tragic. There will be so much needless waste and loss that you will feel furious and heartbroken maybe much of the time. But there will also be comrades, pleasures, and unexpected surprises. There will be allies, not least of which will be the living planet and her web of life.

If you are multisolving and trying to find and stay on that path to a better future, you're unlikely to even be sure that you are on the right path. You won't ever be sure that there's not a better path to be on. There will be collapses and births and the constant need for discernment to figure out which is which. Things have been collapsing for the last many hundred years. Seeds have been being planted for just as long. You will face constant questions. What's worth saving? What should be let go? What can provide a strong foundation on which to build? Which problems might just bundle together in a whole that is easier to heal than its parts?

Only with hindsight, and perhaps not even then, will you know how good your choices were. But you don't need to step into converging crises or multisolving possibilities alone. In these last few pages, in lieu of recipes and roadmaps, I offer you reminders. Reminders of things you've always known, reminders of truths that are your birthright as a complex system alive in a world of systems. Reminders of some of the ideas we've already touched on in this book.

The Multisolving Way

Remember that multisolving, across all of the diverse forms it takes, tends to have in common a core set of attitudes and approaches. I think of this as the Multisolving Way. Across all the layers of the "dandelion" in chapter 9 these approaches come into play. They offer a way of acting in complexity and uncertainty while prioritizing relationships and equity. The Multisolving Way has been weaving itself through all the chapters of this book, but if I had to pull a few elements out here, as we end, I'll name five.

A systems view: embrace interconnection, complexity, learning, and uncertainty. Setting down the illusions of certainty and control, which after all are heavy burdens as well as useless aspirations, there is quite a lot you can carry with you into the uncertainty that is the future. In complex systems, you are never truly alone. One of the tragic effects of the Collection of Objects worldview is that it creates the illusion of separation and isolation. But if you are actively practicing multisolving, designing for emergence, or trying to sharpen your own understanding of a system by listening to others, you're naturally going to be connecting, breaking down silos, and transcending barriers and boundaries. If you are planning for resilience or thinking through scenarios, you're going to be drawn to include others.

You also are not alone in the sense that, around the planet, millions of people are reacting to tumultuous times. Millions like you are trying to understand what could help. Millions of people are activating feedback loops in social movements, developing new technologies to help systems close the gaps between goals and current states, experimenting with new ways of meeting needs. Most of these millions neither you nor I will know directly. But we can trust that we are connected via the many linkages of a complex system.

You will need to test this for yourself, but in my experience enlarging circles of connection and inhabiting a fuller swath of a system brings insight and comfort. I live in an eco-village partly for increased sustainability. But beyond those practical matters, living in this way means I have babies and toddlers to play with even though my own children are grown, as well as friends to swap tomato varieties with me.

Remember also that you are a living system of a living planet. Balancing loops are at work day and night to bring Earth back into balance. Plants in oceans and on land draw down carbon. Whether you are aware of it or not, whether you see it or not, know that many beings are actively steering the complex system that is our world in the same directions that you are trying to steer.

Every time human systems reintegrate, even the tiniest bit, with Earth systems, there is the potential for recovery. I think about this in my canoe. We see loons on almost every lake around where we live. On every river we see bald eagles. When I was growing up in the 1970s, both were rare. The loon population had been decimated by the effects of lead in their waters, and eagles by DDT. When policy changes eliminated both pressures, both populations rebounded.[1]

Much that has been lost in the last few centuries cannot be recovered. Much that is threatened today cannot be saved. Still, much that is vulnerable can be protected. Much that is in decline can recover. Cascades of recovery and regeneration, whether in human communities or in ecosystems, have as much potential for explosive change as do cascades of loss and destruction. The potential for healing is as subject to the rules of unpredictability, thresholds, and nonlinearity as is the potential for collapse.

Every life, every ecosystem, every species is unique and special and sacred. The healing of some, of course, can never make up for the losses that cannot be prevented. But the loss of some does not have to portend the loss of all.

Our inability to control or manage complex systems means that we don't know how much can be saved and how much will be lost. We don't fully understand the potential for healing. What we do know is that our actions steer systems. What we do know is that, all else being equal, the more actions in the direction of healing, the more that will heal. We can't know the outcome, but we can choose our direction.

Remember coherence. Steer change with vision, values, and simple rules. This theme runs throughout the chapters, but it's worth saying one last time. In complex systems focusing too much on whether you are acting in the right place or having the right impact is a distraction. A reinforcing feedback loop, executed well, can take the tiniest of impacts to history-altering proportions. Coherence means that simple rules honed or values cultivated in the humblest situations—at a staff meeting, in your marriage, in your garden—can influence a moment of unforeseen opportunity and shift a system. Multisolving means that the simple act of connecting, with heart and trust, can build power that leads to massive change.

In tumultuous times, many people resist this message. The dangers are so huge and so existential, it can feel that our actions must rise to similar scales. Surely, we need heroes capable of protecting all of Earth with single significant actions. To my mind this is an illusion, one more consistent with the Collection of Objects worldview than with the Web of Relationships worldview. No one has to do everything; we all just have to do something.

Meet the emerging needs that you can. Tend and connect where you are able. And check, relentlessly, for your own alignment with vision, values, and a worldview that is fitting for our intricate and interconnected world. In complex systems, even in tumultuous times—maybe especially in tumultuous times—you may have dozens of tiny opportunities every day to choose how to act, how to intervene, where to connect, and how to connect. Much of your influence will depend on how consistently you conduct yourself and what worldviews and simple rules and visions you embody.

Respond to the moment in a way that also steers the system for the long term. The complexity of systems means every moment is an opportunity to do more than one thing at the same time. You can respond to the latest crisis in a way that both meets the needs of the moment and steers the system for the long term. Perhaps that is the essence of multisolving—doing what needs to be done in the present to keep possibilities open for the future.

The way you take care of people's health can also build political power for the next moment to steer the wider system. You can prepare for rising seas in ways that also restore ecosystems. By embodying a chosen worldview, one that you believe fits how the world works, you can respond to crises and steer systems in ways that build the power of a worldview rising to dominance. How do you prepare food for hungry people after a flood? What is your attitude toward collaboration with others? How are you connected to the farms at the source of your sustenance? When and how will you challenge beliefs about scarcity or competition or domination? How do you embody beliefs about collaboration and connection?

Multisolving rewires systems for moments of need and opportunity. Even after a multisolving project comes to an end, the new relationships it has established remain in the system. When new events arise, the more interconnected system is poised to respond. After multisolving projects conclude, there are also stocks that have been built up. Skills for conversations across sectors or disciplines have improved. Understanding of pressures and incentives in other parts of the system is deeper. Courage and commitment to act are strengthened. We can't always know the impact of rewired systems, but we know that new connections in systems open up new possibilities and new system behaviors.

Prioritize equity. You can disrupt patterns of domination and in their place strengthen patterns of care and collaboration. You have been a complex system since before you were born. You know health when you feel it. You know, for instance, that it would be ridiculous if your body were to declare the left lobe of your liver the supreme boss of all your organs. You know it wouldn't work if all the bloodstream's resources went to the liver and the liver alone. So you know something about ideologies of supremacy and where they lead. You can trust that knowing. It can be a source of courage and validation.

You also are the product of billions of years of evolution, all of which took place in the context of webs of relationship. That's a very long track record compared to some hundreds of years of primacy of the Collections of Objects worldview. Let that knowing also contribute to your courage and confidence.

At the same time, if you, like me, were raised and educated in the dominant culture so steeped in the Collection of Objects worldview, many of your instincts are going to be wrong again and again and again. You have as much to unlearn as you do to learn. As confident as you should be, you should be equally tentative.

That is not to say there are no experts. Though it is not my body of knowledge to share, it is important to acknowledge again that many cultures around the Earth have always and continue to embody a Web of Relationships worldview. Respecting the knowledge systems, ecological knowledge, land rights, and leadership of Indigenous communities is a necessary part of finding our way through tumultuous times.

When multisolving, you are creating a social environment (whether small or large) that is operating from a Web of Relationships worldview. This may be the one place in their lives where the people you bring together can experience what it means to disrupt patterns of domination and seek win-win-win solutions together. Possibilities, once tasted, have a way of changing people, who then carry them into other settings. If Thomas Kuhn's observations hold, every taste of this way of being and working together is a tiny snowball rolling downhill, making it easier for the next experiment and the next one after that.

Foresight: you can prepare for tumultuous times, rapid change, and unexpected opportunity. In all elements of your multisolving you can have one eye on the future. As you cultivate the relationships that make multisolving possible you are also building adaptive capacity, networks that can be drawn on in moments of stress or in moments of opportunity as systems change and shift.

Most of All, Keep Going

In complex systems, the only way to achieve a goal is by moving toward it, constantly swerving off track and then self-correcting. The moth moves up a pheromone gradient by flying until the scent decreases, then turning and flying again until the scent decreases, then turning again. You learned to walk by trying to balance, falling, and trying to balance again. A vine grows toward the light by moving in one direction until there's less light, then changing course.

I know of no other way to navigate converging crises. We just keep going. We cultivate the best defenses we can to protect possibilities for the future that we want. We try things, we learn, we accept feedback from everywhere and everything, and we self-correct, self-correct, self-correct.

We link up in solidarity. That's what multisolving is—lots of little bits of broken and disconnected systems joining hands to heal what is broken. We just act on the truth that the world really is a web of relationships. We trust the web. We learn. We just keep going.

Along the way, we have the great privilege of working with others and of being alive at a time of such consequence. We have the great privilege that, commensurate with the tumultuousness of these times, is the possibility for insight, the possibility for change, the possibility that out of danger and loss might come a way of living that is fit for a living planet.

? Questions for Reflection

- In what ways does knowing more about complex systems help you see possibilities for a just and sustainable future on Earth or the possibilities in your own field or community? Are there ways in which your understanding of complex systems makes the future seem more fragile or tenuous rather than less?
- Now that you've reached the end of the book, what are three things you want to remember or pass on to friends?
- Do you think you will multisolve? Are you already?
- Is this book leading you to think about your work, your goals, or your life differently? What might change? What help might you need?

Acknowledgments

I'm grateful for many teachers, colleagues, and thinking partners who helped me better understand multisolving and how best to communicate about it. I'm especially grateful to systems thinking and system dynamics mentors, including Donella Meadows and John Sterman. I have also learned so much from collaborating on multisolving research and tool development with people like Susanne Moser, Tina Anderson Smith, and Chris Soderquist. With Nathaniel Smith, Tim Palmer, and Angela Park I've had many fruitful conversations that helped clarify the importance of equity and justice in multisolving.

Multisolvers themselves have been generous with their time. They have explained their work and their perspectives in interviews and inspired me with the creativity and impact of what they routinely accomplish.

Moe Bruce and Tina Smith patiently and doggedly reminded me of the importance of finishing this book and sharing it with the world, even when I became busy with other projects.

Work colleagues have been so important in the process of developing the ideas in this book and shaping them into written form. A special thanks to Andrew Jones and colleagues at Climate Interactive who picked up the slack when I took a writing sabbatical to get a start on the book. I am also grateful to my colleagues over the years at Multisolving Institute, including Stephanie McCauley, Kelsi Eccles, and Cassandra Ceballos. They have offered smart feedback, good questions, dedicated work, lighthearted fun, and most of all a huge commitment to supporting multisolvers around the world.

Early on in developing the concepts of multisolving, a few funders had faith in a new idea and helped influence it with rich conversations. Elizabeth Keating of the Why Wait Fund, Mitchell Julis of the Julis-Rabinowitz Family, and Mary Finegan of the Morgan Family Foundation were particularly helpful.

This book has benefitted from the help of editors, especially Michael Metivier and Emily Turner, as well as the whole Island Press team. And it was a pleasure to work with Molly Schafer on the illustrations.

My husband, Phil Rice, has been a steady source of support and a much-valued thinking partner. He's kept home and gardens running well while I've disappeared into my writing, always with good cheer and encouragement. My daughters, Jenna Rice and Nora Rice, and my son-in-law Greg Goedewaagen have encouraged me along the way. And they've shown me their own versions of multisolving as they create lives rooted in community, land, stewardship, art, and food.

Thanks also go to my parents, Ray and Marcia Sawin, for decades of love, support, and encouragement, as well as models of thinking for yourself and working with nature.

Finally, I'm grateful for the land and people of Cobb Hill, the eco-village where I live. I rambled through the woods, paths, and streams while ruminating on parts of this book and always found practical work to do to balance out all the thinking. I couldn't ask for a better training ground about what it means to invest in relationships, learn with others, and try to steer toward collective well-being.

Notes

Introduction. Converging Crises, Cascading Solutions

1. Intergovernmental Panel on Climate Change (IPCC), "Summary for Policy-makers," in *Climate Change 2021—The Physical Science Basis: Working Group I Contribution to the Sixth Assessment Report of the Intergovernmental Panel on Climate Change* (Cambridge: Cambridge University Press, 2023).
2. Intergovernmental Science-Policy Platform on Biodiversity and Ecosystem Services, "Special Report: Nature's Dangerous Decline Is Unprecedented, but It Is Not Too Late to Act," September 28, 2020, http://ipbes.net/news /special-report; Emma Newburger, "Disasters Caused $210 Billion in Damage in 2020, Showing Growing Cost of Climate Change," CNBC, January 7, 2021, https://www.cnbc.com/2021/01/07/climate-change-disasters-cause-210 -billion-in-damage-in-2020.html.
3. Economist Intelligence Unit, "Democracy Index 2020," 2021, https://www.eiu .com/n/campaigns/democracy-index-2020/; Chase Peterson-Withorn, "How Much Money America's Billionaires Have Made During the Covid-19 Pandemic," *Forbes*, April 30, 2021.
4. The COVID Tracking Project, "The COVID Racial Data Tracker," 2020–2021, https://covidtracking.com/race.
5. Brian K. Sullivan, "California Power Grids Strained Amid Worst Heat in 70 Years," *Time*, August 15, 2020, https://time.com/5879916 /california-power-grids-heat-wave/.
6. "One-Third of American Renters Expected to Miss Their August Payment," Bloomberg, August 7, 2020, https://www.bloomberg.com/news /articles/2020-08-07/survey-exposes-america-s-looming-rent-crisis.
7. Annie Nova, "The Pandemic May Cause 40 Million Americans to Lose Their Homes," CNBC, July 30, 2020, https://www.cnbc.com/2020/07/30/what-its -like-to-be-evicted-during-the-coivd-19-pandemic.html.
8. Rick Rojas, "After 2 Hurricanes, Lake Charles Fears Its Cries for Help Have Gone Unheard," *New York Times*, October 20, 2020, https://www.nytimes .com/2020/10/20/us/lake-charles-hurricane-laura-delta.html.
9. Matthew Cappucci, "Hurricane Laura Destroyed a Major Weather Radar as Delta Approaches Louisiana," October 8, 2020, https://www.washingtonpost .com/weather/2020/10/08/radar-lake-charles-hurricane-delta/.
10. "PG&E Failed to Inspect Transmission Lines That Caused Deadly 2018 Wild-fire: State Probe," Reuters, December 3, 2019, https://www.reuters.com /article/us-california-wildfire-pg-e-us-idUSKBN1Y70N8.

11. Vernice Miller-Travis, "Bad News for People Already Overburdened," *Environmental Law Institute* 37, no. 5 (October 2020), https://www.eli.org/the-environmental-forum/bad-news-people-already-overburdened.

12. Angela Park, *Everybody's Movement: Environmental Justice and Climate Change* (Washington, DC: Environmental Support Center, 2009), https://www.adaptationclearinghouse.org/resources/everybody-eys-movement-environmental-justice-and-climate-change.html.

13. Park, *Everybody's Movement*.

14. Donella Meadows, "Dancing with Systems," Donella Meadows Project, 2001, http://donellameadows.org/archives/dancing-with-systems/.

15. Joanna Macy, *The Work That Reconnects* (Gabriola Island, BC: New Society Publishers, 2006).

16. Climate Justice Alliance, "The Principles of Environmental Justice: Climate Justice Alliance," 2020, https://climatejusticealliance.org/ej-principles/.

17. Robert D. Bullard, Glenn S. Johnson, and Angel O. Torres, *Environmental Health and Racial Equity in the United States: Building Environmentally Just, Sustainable, and Livable Communities* (Washington, DC: American Public Health Association, 2011).

Chapter One. Multisolving: Promises and Obstacles

1. Elizabeth Sawin, "The Magic of 'Multisolving,'" July 18, 2018, https://ssir.org/articles/entry/the_magic_of_multisolving#.

2. Anuradha Varanasi, "What Are the Hidden Co-Benefits of Green Infrastructure?" *State of the Planet* (blog), September 3, 2019, https://news.climate.columbia.edu/2019/09/03/hidden-benefits-green-infrastructure/.

3. Paul R. Epstein, Jonathan J. Buonocore, Kevin Eckerle, Michael Hendryx, Benjamin M. Stout III, Richard Heinberg, Richard W. Clapp, et al., "Full Cost Accounting for the Life Cycle of Coal," *Annals of the New York Academy of Sciences* 1219, no. 1 (February 2011): 73–98. https://doi.org/10.1111/j.1749-6632.2010.05890.x.

4. Wendell Berry, "Solving for Pattern," in *The Gift of Good Land: Further Essays, Cultural and Agricultural* (Berkeley, CA: Counterpoint, 1981), 137.

5. Toby Hemenway, "Toby Hemenway: Explaining Permaculture," *Resilient Life Podcast*, May 14, 2014, https://www.peakprosperity.com/toby-hemenway-explaining-permaculture/.

6. Gregory Cajete, "Plant, Food, Medicine, and Gardening," in *Native Science: Natural Laws of Interdependence* (Santa Fe, NM: Clear Light Publishers, 2000), 142–43.

7. Ed Reed, "Nigeria Plans Solar Household Boost," Energy Voice, June 12, 2020, https://www.energyvoice.com/renewables-energy-transition/245447/nigeria-plans-solar-household-boost/.

8. Anastasia Moloney, "Colombia's Medellin Pushes 'Eco-City' Aims in Coronavirus Recovery," Reuters, May 26, 2020, https://www.reuters.com/article/us-health-coronavirus-colombia-climate-c-idUSKBN2321XZ.

9. Alan Kohll, "New Study: Air Quality and Natural Light Have the Biggest Impact on Employee Well-Being," *Forbes*, August 13, 2019, https://www

.forbes.com/sites/alankohll/2019/08/13/new-study-air-quality-and-natural
-light-have-the-biggest-impact-on-employee-well-being/.

10. Elizabeth Howton, "Nearly Half the World Lives on Less than $5.50 a Day,"
World Bank, October 17, 2018, https://www.worldbank.org/en/news
/press-release/2018/10/17/nearly-half-the-world-lives-on-less-than-550-a-day.

11. World Bank, "Indigenous Peoples," March 19, 2021, https://www.worldbank
.org/en/topic/indigenouspeoples.

12. World Health Organization, *COP26 Special Report on Climate Change and
Health: The Health Argument for Climate Action* (Geneva: World Health
Organization, 2021), https://www.who.int/publications-detail-redirect
/9789240036727.

Chapter Two. Stocks

1. Donella H. Meadows and Diana Wright, *Thinking in Systems: A Primer* (White
River Junction, VT: Chelsea Green, 2008), 96.

2. Brad Plumer, Nadja Popovich, and Brian Palmer, "How Decades of Racist
Housing Policy Left Neighborhoods Sweltering," *New York Times*, August 31,
2020, https://www.nytimes.com/interactive/2020/08/24/climate
/racism-redlining-cities-global-warming.html.

3. Climate Justice Alliance, "Just Transition: A Framework for Change," 2019,
https://climatejusticealliance.org/just-transition/.

4. Brian Kahn, "What's the Best Way to Decarbonize Your Home?" Gizmodo,
August 20, 2021, https://gizmodo.com/what-s-the-best-way-to
-decarbonize-your-home-1847518817.

5. Rakesh Kochhar and Mohamad Moslimani, "2. Wealth Gaps across Racial and
Ethnic Groups," *Pew Research Center Race and Ethnicity* (blog), December 4,
2023, https://www.pewresearch.org/race-ethnicity/2023/12/04
/wealth-gaps-across-racial-and-ethnic-groups/.

Chapter Three. Flows

1. Ben Goldfarb, "How Beavers Became North America's Best Firefighter,"
National Geographic, September 22, 2020, https://www.nationalgeographic
.com/animals/article/beavers-firefighters-wildfires-california-oregon.

2. Katelyn Newman, "Portland, Oregon, Taxes Big Businesses for Clean Energy
Efforts," *US News and World Report*, January 30, 2019, https://www.usnews
.com/news/cities/articles/2019-01-30/portland-oregon-taxes-big
-businesses-to-fund-clean-energy-jobs.

Chapter Four. Reinforcing Feedback

1. Raquel Vaquer-Sunyer and Carlos M. Duarte, "Thresholds of Hypoxia for
Marine Biodiversity," *Proceedings of the National Academy of Sciences* 105, no.
40 (October 7, 2008): 15452–57, https://doi.org/10.1073/pnas.0803833105.

2. United Nations High Commissioner for Refugees, "UNHCR—Refugee Statis-
tics," UNHCR, June 18, 2021, https://www.unhcr.org/refugee-statistics/.

3. Hannah Ritchie and Max Roser, "Plastic Pollution," Our World in Data, September 1, 2018, https://ourworldindata.org/plastic-pollution; United Nations,

"Leaving No One Behind: The United Nations World Water Development Report 2019," March 18, 2019, https://www.unwater.org/publications /world-water-development-report-2019/.

4. Geoffrey M. Cooper, "The Development and Causes of Cancer," in *The Cell: A Molecular Approach* 2nd ed. (Sunderland, MA: Sinauer Associates, 2000), https://www.ncbi.nlm.nih.gov/books/NBK9963/.

5. Ezra Klein, "Transcript: Ezra Klein Interviews Heather McGhee about the Cost of Racism," *New York Times*, February 16, 2021, https://www.nytimes .com/2021/02/16/podcasts/ezra-klein-podcast-mcghee-transcript.html.

6. Jillian Ambrose, "Energy Firms Urged to Mothball Coal Plants as Cost of Solar Tumbles," *Guardian*, June 2, 2020, https://www.theguardian.com /business/2020/jun/02/energy-firms-urged-to-mothball-coal-plants -as-cost-of-solar-tumbles.

Chapter Five. Balancing Feedback

1. Espigoladors, "We Fight to Stop Food Waste and Losses," https:// espigoladors.cat/en/.

2. Espigoladors, "We Fight to Stop Food Waste and Losses."

3. Environmental Protection Agency, "Overview of the Clean Air Act and Air Pollution," Collections and Lists, February 27, 2015, https://www.epa.gov /clean-air-act-overview.

4. Department of Economic and Social Affairs, "The 17 Goals," United Nations, https://sdgs.un.org/goals.

5. "New Member Resources: Saving Energy through Behavior Change," Global Green and Healthy Hospitals, 2017, https://www.greenhospitals.net /new-member-resources-saving-energy-through-behavior-change/.

6. UNEP and UNEP Copenhagen Climate Centre, *Emissions Gap Report 2020* (Nairobi: UN Environment Programme, December 1, 2020), http://www.unep .org/emissions-gap-report-2020.

Chapter Six. The Behavior of Whole Systems

1. Naomi Oreskes and Erik M. Conway, *Merchants of Doubt: How a Handful of Scientists Obscured the Truth on Issues from Tobacco Smoke to Global Warming* (New York: Bloomsbury, 2011).

2. James Baldwin, "Black English: A Dishonest Argument," in *The Cross of Redemption: Uncollected Writings*, edited by Randall Kenan (New York: Vintage, 2011).

3. Michael Goodman, "Systems Thinking as a Language," *Systems Thinker*, November 12, 2015, https://thesystemsthinker.com/systems-thinking -as-a-language/; Donella Meadows, "Envisioning a Sustainable World," *Solutions Journal* (blog), February 21, 2016, https://thesolutionsjournal.com /2016/02/22/envisioning-a-sustainable-world/; Nathaniel Smith, "History Is Present in the Voting Booth: The Making of a Values Revolution," *Nonprofit Quarterly*, November 17, 2020, https://nonprofitquarterly.org/history-is -present-in-the-voting-booth-the-making-of-a-values-revolution/; Royce Holladay and Mallary Tytel, *Simple Rules: A Radical Inquiry into Self: Going*

beyond Self-Help to Generate Self-Hope (Apache Junction, AZ: Gold Canyon Press, 2011); "Simple Rules," Human Systems Dynamics Institute, https://www.hsdinstitute.org/about-hsd-institute/simple-rules.html.

Chapter Seven. Rising to the Challenge of Complex Systems

1. Donella Meadows, "Dancing with Systems," Donella Meadows Project, 2001, http://donellameadows.org/archives/dancing-with-systems/.
2. Peter Senge, The Fifth Discipline: The Art and Practice of the Learning Organization (New York: Crown, 2010).
3. D. Kriebel, J. Tickner, P. Epstein, J. Lemons, R. Levins, E. L. Loechler, M. Quinn, R. Rudel, T. Schettler, and M. Stoto, "The Precautionary Principle in Environmental Science," Environmental Health Perspectives 109, no. 9 (September 2001): 871–76, https://doi.org/10.1289/ehp.01109871.
4. Kriebel et al., "The Precautionary Principle."
5. Ezra Klein, "Interview with Heather McGhee." The Ezra Klein Show, podcast transcript, February 16, 2021, https://www.nytimes.com/2021/02/16/podcasts/ezra-klein-podcast-mcghee-transcript.html.
6. Vivek Ravichandran and Sacoby Wilson, "Utilizing Community-Based Participatory Research (CBPR) Methods to Develop Hyperlocal Low-Cost Real-Time Air Quality Monitors in Cheverly, Maryland," American Meteorological Society, 2023, https://ams.confex.com/ams/103ANNUAL/meetingapp.cgi/Paper/415031.
7. Na'Taki Osborne Jelks, Timothy L. Hawthorne, Dajun Dai, Christina H. Fuller, and Christine Stauber, "Mapping the Hidden Hazards: Community-Led Spatial Data Collection of Street-Level Environmental Stressors in a Degraded, Urban Watershed," International Journal of Environmental Research and Public Health 15, no. 4 (April 2018): 825, https://doi.org/10.3390/ijerph15040825.
8. Maya Lin Studio, "What Is Missing?" accessed March 11, 2024, https://www.mayalinstudio.com/memory-works/what-is-missing.
9. Holland Cotter, "In Maya Lin's 'Ghost Forest,' the Trees Are Talking Back," New York Times, July 1, 2021, https://www.nytimes.com/2021/07/01/arts/design/maya-lin-ghost-forest.html.
10. Daniel Kahneman and Angus Deaton, "High Income Improves Evaluation of Life but Not Emotional Well-Being," Proceedings of the National Academy of Sciences 107, no. 38 (September 21, 2010): 16489–93, https://doi.org/10.1073/pnas.1011492107.
11. Donella Meadows, "We Can't Keep Stealing from the Future: A Sustainable Economy—One That's Efficient and Sufficient—Doesn't Have to Mean Sacrifice," Los Angeles Times, April 27, 1992, https://www.latimes.com/archives/la-xpm-1992-04-27-me-551-story.html.
12. Rhiana Gunn-Wright, "Opinion: Think This Pandemic Is Bad? We Have Another Crisis Coming," New York Times, April 15, 2020, https://www.nytimes.com/2020/04/15/opinion/sunday/climate-change-covid-economy.html.
13. "How the Green New Deal Changed the Conversation," New Republic, November 3, 2022, https://newrepublic.com/article/167994/green-new-deal-changed-conversation.

Chapter Eight. Steering Systems

1. Donella Meadows, "Envisioning a Sustainable World," *Environmental Conservation* 23, no. 4 (December 1996): 117–29, https://doi.org/10.1017/S0376892900039448.

2. Peter M. Senge, *The Fifth Discipline: The Art and Practice of the Learning Organization* (New York: Doubleday/Currency, 2006), http://archive.org/details/fifthdisciplinea0000seng_k2n0.

3. "Interview With Carolyn Raffensperger On The Revolutionary Idea Of Putting Safety First | Interviews by Derrick Jensen," November 1, 2002. https://derrickjensen.org/2002/11/interview-carolyn-raffensperger-putting-safety-first/.

4. "Simple Rules," Human Systems Dynamics Institute, https://www.hsdinstitute.org/resources/simple-rules.html.

5. Annie Gowen, "The Town That Built Back Green," *Washington Post*, October 22, 2020, https://www.washingtonpost.com/climate-solutions/2020/10/22/greensburg-kansas-wind-power-carbon-emissions/.

6. Donella Meadows, "Leverage Points: Places to Intervene in a System," *Solutions Journal* (blog), February 21, 2016, https://thesolutionsjournal.com/2016/02/22/leverage-points-places-to-intervene-in-a-system/.

7. Thomas Berry, *Evening Thoughts: Reflecting on Earth as Sacred Community* (San Francisco: Sierra Club Books, 2006), 17–18, http://archive.org/details/eveningthoughtsr0000berr.

8. Carl Anthony, *The Earth, the City, and the Hidden Narrative of Race* (New York: New Village Press, 2017), 186.

9. Gregory Cajete, *Native Science: Natural Laws of Interdependence* (Santa Fe, NM: Clear Light Publishers, 2000), 75, http://archive.org/details/nativesciencenat0001caje.

10. Kyle Whyte, "Critical Investigations of Resilience: A Brief Introduction to Indigenous Environmental Studies and Sciences," *Daedalus* 147, no. 2 (2018): 136–47, https://doi.org/10.1162/DAED_a_00497.

11. Joanna Macy, *Mutual Causality in Buddhism and General Systems Theory: The Dharma of Natural Systems* (Albany: State University of New York Press, 1991), 19, http://archive.org/details/mutualcausalityi0000macy.

12. Farhana Sultana, "The Unbearable Heaviness of Climate Coloniality," *Political Geography* 99 (November 1, 2022): 102638, https://doi.org/10.1016/j.polgeo.2022.102638.

13. Farhana Sultana, "Whose Growth in Whose Planetary Boundaries? Decolonising Planetary Justice in the Anthropocene," *Geo: Geography and Environment* 10, no. 2 (2023): e00128, https://doi.org/10.1002/geo2.128.

14. Thomas S. Kuhn, *The Structure of Scientific Revolutions* (Chicago: University of Chicago Press, 1970), http://archive.org/details/structureofscien00kuhnrich.

15. Julia Manchester, "Majority of Young Adults in US Hold Negative View of Capitalism: Poll," *The Hill* (blog), June 28, 2021, https://thehill.com/homenews/campaign/560493-majority-of-young-adults-in-us-hold-negative-view-of-capitalism-poll/.

Chapter Nine. Multisolving in Action

1. "Multisolving at the Intersection of Health and Climate," Multisolving Institute, https://www.multisolving.org/resources /multisolving-at-the-intersection-of-health-and-climate/.
2. "Multisolving for Climate Resilience," Multisolving Institute, https://www .multisolving.org/resources/multisovling-for-climate-resilience/.
3. "Multisolving for Climate Resilience."
4. "Just Communities: The Evolution of EcoDistricts," Just Communities, https:// justcommunities.info/.
5. "Manifesto for Just Communities," Just Communities, https:// justcommunities.info/about/manifesto/.
6. "Multisolving for Climate Resilience."
7. Reina Sultan and Micah Herskind, "What Is Abolition, and Why Do We Need It?" Transform Harm, July 24, 2020, https://transformharm.org/ab_resource /what-is-abolition-and-why-do-we-need-it/.
8. David P. Barash, "Costa Rica's Peace Dividend: How Abolishing the Military Paid Off," *Los Angeles Times*, December 15, 2013, https://www.latimes.com /opinion/la-xpm-2013-dec-15-la-oe-barash-costa-rica-demilitarization -20131208-story.html.
9. Elizabeth R. Sawin, Kelsi Eccles, Susanne Moser, and Tina A. Smith, "Multi-solving: Making Systems Whole, Healthy, and Sustainable," *Stanford Social Innovation Review*, November 1, 2023, https://ssir.org/articles/entry /multisolving_making_systems_whole_healthy_and_sustainable.
10. Elizabeth Sawin, Nathaniel Smith, and Tina Anderson Smith, "Equity, Health, Resilience, and Jobs: Lessons from the Just Growth Circle," *Nonprofit Quarterly*, August 22, 2019, https://nonprofitquarterly.org /equity-health-resilience-and-jobs-lessons-from-the-just-growth-circle/.
11. Drew Kann, "Atlanta, Nonprofits Getting Millions for Urban Tree Protection," *Atlanta Journal-Constitution*, September 14, 2023, https://www.ajc .com/news/atlanta-nonprofits-getting-millions-for-urban-tree-protection /IMNV5EADOVCXLFZQLLK2YHNMSY/.
12. Christine McGowan, "A Vermonter's Approach to Sustainable, Affordable Housing," Vermont Sustainable Jobs Fund, April 29, 2021, https://www.vsjf .org/2021/04/29/a-vermonters-approach-to-sustainable-affordable-housing/.
13. Mary Williams Engisch, "Reflecting on Tropical Storm Irene Following This Month's Historic Floods," Vermont Public, July 26, 2023, https://www.vermontpublic.org/local-news/2023-07-26 /reflecting-on-tropical-storm-irene-following-this-months-historic-floods.
14. Engisch, "Reflecting on Tropical Storm Irene."
15. adrienne maree brown, *Emergent Strategy: Shaping Change, Changing Worlds* (Chico, CA: AK Press, 2017), 42.
16. Isabelle Anguelovski, James J. T. Connolly, Hamil Pearsall, Galia Shokry, Melissa Checker, Juliana Maantay, Kenneth Gould, Tammy Lewis, Andrew Maroko, and J. Timmons Roberts, "Why Green 'Climate Gentrification' Threatens Poor and Vulnerable Populations," *Proceedings of the National Academy of Sciences* 116, no. 52 (December 26, 2019): 26139–43, https://doi.org/10.1073/pnas.1920490117.

Chapter Ten. Multisolving and Equity

1. "Glossary of Terms," Agency for Toxic Substances and Disease Registry, November 13, 2018, https://www.atsdr.cdc.gov/glossary.html#G-A-.
2. Betsy Reid, "The Voice of the African American Male in Philanthropy: A Conversation on Power, Courage, Agency, and Representation," *PEAK Grantmaking* (blog), June 19, 2020, https://www.peakgrantmaking.org/insights /the-voice-of-the-african-american-male-in-philanthropy-a-conversation -on-power-courage-agency-and-representation/.
3. Judd Kessler and Corinne Low, "Research: How Companies Committed to Diverse Hiring Still Fail," *Harvard Business Review*, February 11, 2021, https://hbr.org/2021/02/research-how-companies-committed -to-diverse-hiring-still-fail.
4. Heather C. McGhee, *The Sum of Us: What Racism Costs Everyone and How We Can Prosper Together* (New York: One World, 2021).
5. "Transcript: Ezra Klein Interviews Heather McGhee about the Cost of Racism," *New York Times*, February 16, 2021, https://www.nytimes .com/2021/02/16/podcasts/ezra-klein-podcast-mcghee-transcript.html.
6. Isabelle Anguelovski, James J. T. Connolly, Hamil Pearsall, Galia Shokry, Melissa Checker, Juliana Maantay, Kenneth Gould, Tammy Lewis, Andrew Maroko, and J. Timmons Roberts, "Why Green 'Climate Gentrification' Threatens Poor and Vulnerable Populations," *Proceedings of the National Academy of Sciences* 116, no. 52 (December 26, 2019): 26139–43, https://doi .org/10.1073/pnas.1920490117.
7. Mark Moran, "Public Health Physician Offers Vivid Parables of Social Inequity," *Psychiatric News* 53, no. 21 (November 2, 2018), https://doi.org/10.1176 /appi.pn.2018.11a5.
8. Whitney Pfeifer, "From 'Nothing About Us Without Us' to 'Nothing Without Us,'" National Democratic Institute, March 28, 2022, https://www.ndi.org /our-stories/nothing-about-us-without-us-nothing-without-us.
9. James Charlton, *Nothing About Us Without Us: Disability Oppression and Empowerment* (Berkeley: University of California Press, 1998), 3.
10. Daniel Kim, "Success to the Successful: Self-Fulfilling Prophecies," *Systems Thinker* (blog), February 22, 2016, https://thesystemsthinker.com /success-to-the-successful-self-fulfilling-prophecies/.

Chapter Eleven. Multisolving in Tumultuous Times

1. Camila Domonoske, "Why Car Prices Are Still So High—and Why They Are Unlikely to Fall Anytime Soon," NPR, March 18, 2023, https://www.npr.org /2023/03/18/1163278082/car-prices-used-cars-electric-vehicles-pandemic.
2. Kent Wainscott, "Milwaukee's Deep Tunnel Open for 20 Years," WISN, August 9, 2013, https://www.wisn.com/article/milwaukee-s-deep -tunnel-open-for-20-years/6316840.
3. Elliot Smith, "Record Flooding Hammers the African Sahel, the Latest in a Series of Shocks," CNBC, September 10, 2020, https://www.cnbc.com /2020/09/10/record-flooding-hammers-the-african-sahel-the-latest-in -a-series-of-shocks.html.

4. Rebecca Lindsey, "Intense Drought in the U.S. Southwest Persisted throughout 2018, Lingers into the New Year," NOAA, February 6, 2019, http://www.climate.gov/news-features/featured-images /intense-drought-us-southwest-persisted-throughout-2018-lingers-new.

5. Willy C. Shih, "Global Supply Chains in a Post-Pandemic World," *Harvard Business Review*, September 1, 2020, https://hbr.org/2020/09 /global-supply-chains-in-a-post-pandemic-world.

6. Elizabeth Weise, "Extreme Heat Already Disrupts Air Travel. With Climate Change, It's Going to Get Worse," Science X, March 28, 2022, https://phys.org /news/2022-03-extreme-disrupts-air-climate-worse.html.

7. Lora Cecere, "The Lessons from the Lonely Roll of Toilet Paper in the Pandemic," *Forbes*, April 21, 2021, https://www.forbes.com/sites /loracecere/2021/04/21/the-lessons-from-the-lonely-roll-of -toilet-paper-in-the-pandemic/.

8. A. R. Siders and Katharine Mach, "'Managed Retreat' Done Right Can Reinvent Cities So They're Better for Everyone—and Avoid Harm from Flooding, Heat and Fires," *Conversation*, June 21, 2021, http://theconversation.com /managed-retreat-done-right-can-reinvent-cities-so-theyre-better-for -everyone-and-avoid-harm-from-flooding-heat-and-fires-163052.

Chapter Twelve. Going Forth

1. Kayla Webley, "Bald Eagle Leaves Endangered Species List," NPR, June 28, 2007, https://www.npr.org/2007/06/28/11504430/bald-eagle-leaves -endangered-species-list; Eric W. Hanson and Doug Morin, "The 2021 Breeding Status of Common Loons in Vermont," Vermont Center for Ecosystem Studies, 2021, https://vtecostudies.org/wp-content/uploads/2022/01 /Vermont-Loon-Report-2021.pdf.

Index

Note: page numbers followed by f and t refer to figures and tables, respectively.

About the Author

Elizabeth Sawin, PhD, has dedicated her career to the theory and practice of creating change in complex systems. She trained in system dynamics computer simulation with Donella Meadows at the Sustainability Institute. At the Institute she also supported sustainability leaders from around world as they used system approaches to conserve land, enact climate policy, restore rivers, and pro-

Photo credit: Jenna Rice

mote healthy communities. In 2010 Beth cofounded the think tank Climate Interactive to create tools for grappling with the complexity of the climate system. She also developed the concept of multisolving to describe a style of leadership she observed around the world where people collaborate to achieve multiple goals, like climate, equity, and health, at the same time. In 2021 Beth founded Multisolving Institute to share and extend this research and to develop tools tailored for multisolving. Her work has been widely covered, including in the *New York Times* and the *Washington Post*, and she writes and speaks internationally about leadership in complex systems and multisolving. Beth holds a PhD from MIT.